William O'Connor Morris

The French Revolution and First Empire

An Historical Sketch

William O'Connor Morris

The French Revolution and First Empire
An Historical Sketch

ISBN/EAN: 9783337230135

Printed in Europe, USA, Canada, Australia, Japan

Cover: Foto ©berggeist007 / pixelio.de

More available books at **www.hansebooks.com**

THE

FRENCH REVOLUTION

AND FIRST EMPIRE:

AN HISTORICAL SKETCH.

BY

WILLIAM O'CONNOR MORRIS,

SOMETIME SCHOLAR OF ORIEL COLLEGE, OXFORD.

καὶ μὴν ἔργῳ κοὐκ ἔτι μύθῳ
χθὼν σεσάλευται·
βρυχία δ' ἠχὼ παραμυκᾶται
βροντῆς, ἕλικες δ' ἐκλάμπουσι
στεροπῆς ζάπυροι,
στρόμβοι δὲ κόνιν εἰλίσσουσι.
                                    Æschylus.

LONDON:

LONGMANS, GREEN, AND CO.

1874.

Dedicated to

HENRY REEVE, ESQ., D.C.L.

AS A MARK OF THE REGARD AND ESTEEM OF

THE AUTHOR.

PREFACE.

THIS little work was intended to be a number
of the 'Epochs of History,' in course of publication by the Messrs. Longman. It now
appears, however, in a separate form, the
Editor of 'Epochs of History' having considered it suited to readers more advanced in
years than those for whom that series is especially designed.

It is unnecessary to say that, in an epitome
of this kind, innumerable details must be altogether left out, and that a small space only
can be allotted to even important occurrences
which would be set forth at length in a complete narrative. Nor has it been possible for
me—my object being to describe the principal
facts of the French Revolution and First Empire—to comment largely on the institutions of

old France, or to show fully how they contributed to the events that followed 1789. An
abridgment cannot be a real History; and,
apart from defects peculiar to it, I am con-
scious that this volume must at best be an
imperfect miniature of the grand drama of
human action and life which it endeavours to
delineate. Still I am not without hope that I have
represented, in something like exact outline,
the great features of that period of trouble and
war through which France passed from 1789
to 1815 ; and I trust that I have placed events
in their true proportions, and that the opinions I
have expressed are correct and moderate. The
present time, it must be allowed, is favourable
for a publication of this kind, even though it pur-
ports to be only a sketch. French and English
literature has of late years teemed with documents of the greatest value on the Revolution
and Napoleon I., and I have carefully studied
most of these sources of information. The
events, too, of the war of 1870 bring again
before our eyes what the Emperor achieved in
the field, though victory has shifted from the
standards of one race to those of another; the
national defence of France in 1871 reflects

light on that of 1793 ; and in the crimes and madness of the lately suppressed Commune of Paris, we see an image of the Reign of Terror.

It should be added that this volume is, in my judgment, not unfitted for comparatively young students at least, though now offered to the general public. I have but slightly altered the original text; but I have introduced a few notes, and referred to some authorities, for the most part easily accessible ; and this I hope will be of use to the ordinary reader.

DUBLIN : *February* 3, 1874.

CONTENTS.

CHAPTER I.

STATE OF FRANCE BEFORE THE REVOLUTION.

PAGE

General character and results of the French Revolution . 1
It exhibited vividly the peculiar features of the French
national character 2
The Monarchy of France before the Revolution; abuses in
the system of Government 4
The Church and the Nobility. Why these orders had be-
come generally disliked 7
The power of the ruling orders was divided and decayed,
and consequently weak 9
Dissensions between the Crown, the Church, and the Nobles,
and discredit of all authority 11
State of the Commons of France. The middle classes; how
they were cut off from the People, and the results . . 13
Condition of the Peasantry 15
Of the population of the towns 17
Destructive and sceptical tone of contemporary French
Literature 17
The progress of evil continued down to the Revolution . 19
Brilliant anticipations of the future in France . . . 21

CHAPTER II.

THE STATES-GENERAL AND NATIONAL ASSEMBLY.

1789 Meeting of the States-General 23
The Commons declare themselves the National Assembly . 24
The oath of the Tennis Court, June 20 26

PAGE

Dismission of Necker, and formation of a reactionary
 Ministry 29
Rising in Paris 29
Mutiny of the French Guards and insubordination in the
 Army 30
The Commúne, the National Guard 32
Siege and storming of the Bastille, July 14, 1789 . . 33
Beginning of the emigration of the Nobles 34
Recall of Necker 35
The King sanctions what had been done in Paris . . 35
The Tricolour Flag 35
Risings in the Provinces 36
Legislative measures of the Assembly 38
Sudden abolition of the feudal burdens, August 4, 1789 . 39
October 5 and 6, 1789 40
The King and Royal Family taken to Paris from Versailles 42
Growing power of the rabble of Paris 43

CHAPTER III.

THE CONSTITUTION OF 1790–1.

Character of the period from the autumn of 1789 to the
 summer of 1791 44
The National Assembly begins to frame the Constitution . 46
Useful reforms 46
Wild and precipitate innovations 47
The Rights of Man 47
Confiscation of Church and corporate property. Abolition
 of tithes 47
Equality 48
Provinces transformed into Departments 48
Character of the new Constitution; its glaring defects . 48
Administrative measures 50
Assignats 50
False system of taxation 51
Undue favour shown to Paris 51
The feast of the Federation, July 14, 1790. Enthusiasm in
 Europe 52
Evil consequences of innovation; signs of disorder and
 anarchy 53

PAGE

The Jacobin and other clubs 55
Weakness of Conservative elements in the Assembly . . . 56
Attempts of Mirabeau to check the disorganisation of the
   State 58
His death 59
Threatening attitude of Foreign Powers 59
Dangerous projects of the King and Queen . . . 60
The flight to Varennes, June 20, 1791 61
Consequences of this event 62

CHAPTER IV.

THE LEGISLATIVE ASSEMBLY.

1791 Character of the period from the summer of 1791 to August
   10, 1792 65
Deceptive calm. Meeting of the new Legislative Assembly. 66
Character of this Body 67
The Gironde 67
Dissensions between the King and the Assembly . . 67
Disorders in the Provinces 68
Religious and social troubles 69
Massacre at Avignon (Oct. 16–20, 1791) 69
Louis directly opposes the Assembly 70
Nov. 1791 70
Indignation of the Assembly 71
Attitude of the Commune of Paris. The Clubs and dema-
   gogues 71
Menaces of Foreign Powers 72
Declaration of Pilnitz, August, 1791 72
Projects of King and Queen 73
Divisions in the Assembly 73
It declares war against Austria. Prussia joins Austria. 74
French failures in Belgium. April, May, 1792 . . . 74
Indignation in France 74
1791- Insensate conduct of Louis 75
1792 He resists the decrees of the Assembly, and dismisses the
   Gironde ministry and Dumouriez 76
Outbreak of June 20, 1792 76

PAGE

1792 Encouraged by the popular leaders in the Assembly and by
    the Commune of Paris 77
    Reaction in favour of Louis 78
    New efforts of the Demagogues and the Commune . . 79
    Proclamation of Brunswick 80
    Invasion of France by the Prussians and Austrians, July
      and August, 1792 81
    The Assembly paralyzed; power passes to the Demagogues 82
    Preparations for a rising. Danton 82
    Paris on the night of August 9, 1792 83
    Attitude of the King and the Court 84
    August 10, 1792 85
    The armed populace at the Tuileries 86
    The King and Royal Family take refuge in the Assembly . 86
    The Tuileries attacked and pillaged. Massacre of the Swiss 87
    Reflections on the rising of August 10 88

CHAPTER V.

THE CONVENTION, TO THE FALL OF THE MODERATES.

Effects of August 10 90
The Convention summoned 90
The King and Royal Family imprisoned in the Temple . 90
Violent measures of the Commune of Paris . . . 91
Lafayette throws up his command; advance of the German
    armies to Verdun 92
Massacre of September 93
September 2 to 6, 1792 93
Frightful scenes in Paris 94
The Assembly indignant at the massacre 95
Battle of Valmy. Its great results; retreat of the Prussians
    and Austrians 96
Meeting of the Convention, September 22, 1792 . . . 97
Parties in it 97
France declared a Republic, September 22, 1792 . . 97
Offer of liberty to foreign nations, November 19 . . 98
Dissension renewed between the Moderates and Jacobins . 98
Trial of Louis XVI. December 11, 1792 99
Sentence of death pronounced by a majority of one . . 100
Execution of Louis XVI. January 21, 1793 . . . 101

PAGE

1793 Reflections on this event 103
Character and conduct of the King 104
Coalition of Europe against France . . . . .⟨104⟩
Battle of Jemmapes and early successes of the French, No-
vember-December 1792, January-February 1793 . . 105
Fierce struggle of parties in France 108
The Gironde denounced by the Jacobins and Demagogues . 108
Distress and social disorders . . . . . ⟨109⟩
Advance of the Coalition . . . . . ' . . 109
Battle of Neerwinden, March 18, 1793 109
Flight of Dumouriez. He throws up his command . . 110
Increasing power of the Jacobins. Danton. His energy .⟨110⟩
Formation of the Committee of Public Safety, April 6, 1793 111
Violent measures proposed by Danton 111
And decreed by Convention 112
Propositions of the Commune of Paris 112
The party of violence generally prevails 112
The Gironde and Moderates denounce extreme measures . 113
The Commission of Twelve 113
The forces of Anarchy become supreme . . . .⟨114⟩
Danton tries in vain to reconcile the contending parties . 114
Death-struggle between the Moderates and Jacobins . . 114
Rising of May 31 and June 2, 1793 114
Fall of the Moderates 115
Reflections on this event . . . .⟨115⟩

CHAPTER VI.

THE REIGN OF TERROR. WAR. FALL OF ROBESPIERRE.

1793 Triumph of the party of violence in the Convention . . 117
Risings against it in the Provinces. Civil war . . . 117
Beginning of the war of La Vendée 118
Energetic measures of Danton and the leaders in power . 119
The levée en masse 119
The Constitution of 1793 . . . . . . ⟨120⟩
The maximum. The Revolutionary Tribunal . . . 120
The risings in part of France quelled 120
Feebleness of the Coalition . . . . . · . 121
Waste of time, in action and dissensions 122

PAGE

1793   Sept. 8, 1793 123
        The Republic successful at home and abroad . . . 123
        Carnot. Hoche 123
        Battle of Wattignies, October 16, 1793 124
        Fall of Lyons, October 9, 1793 125
        Great-defeat of the Vendeans at Savenay, December 23, 1793 125
        Siege and fall of Toulon, December 19, 1793 . . . 125
        First appearance on the scene of Napoleon Bonaparte . . 126
        The Reign of Terror 126
        The Convention, a mere instrument of the Jacobin leaders . 127
        The Committee of Public Safety all powerful . . . 127
        Its tyranny and terrible expedients 127
        Wild social changes 128
        Atheism declared truth by the Commune of Paris . . 128
        Appearance of Paris during the Reign of Terror . . . 129
        The levies hurried to the frontier 130
        Appearance of the Convention 130
        Dissolution of society and of morality 131
        Cruelty and suspicions of populace 131
        General licentiousness 132
        The Goddess of Reason at Notre-Dame . . . 132
        Scenes in the prisons 133
        The Revolutionary Tribunal and its work 134
        Trial and execution of Marie Antoinette, October 14-16, 1793 136
        Her character and conduct 137
        Divisions among the Jacobin rulers 138
        Three factions form themselves 138
        Growing ascendency of Robespierre 139
        He becomes supreme in the State 140
1794   Destruction of Hébert and the leaders of the Commune, of
            Danton and his followers, March 24, April 3, 1794 . . 140
        Character of Danton 141
        Dictatorship of Robespierre 141
        His measures to secure his power 141
        The worship of the Supreme 141
        Terror at its height 141
        Frightful state of Paris 142
        Massacres in the Provinces 143
        The Republic obtains fresh successes in the campaign of
            1794 144

PAGE

1794 English naval victory of June 1 145
The allies defeated on all other points of the theatre . . 145
Battle of Fleurus, June 26, 1794 145
Reaction against the Reign of Terror 146
Fall of Robespierre, July 27, 1794 147
Execution of Robespierre, St. Just, Couthon, and others,
July 28, 1794 147
Reflections on this event 148
The Terrorists were not able men 148
Notwithstanding the efforts of the French, the Allies could
have put down the Revolution 149

CHAPTER VII.

THERMIDOR. FRENCH CONQUESTS.

Reaction of Thermidor 151
The prisons opened 151
Punishment of several of the Terrorists 151
Abolition of the Revolutionary Tribunal, and of the maxi-
mum 152
The forcing of the value of assignats discontinued . . 152
The populace of Paris kept down. The Jacobin Club sup-
pressed 152
Violence of the Reaction 153
Revolution in manners 154
The Jeunesse Dorée . , ... 155
Renewed troubles 155
Scarcity and distress 156
The Jacobin party tries to rally 157
1795 Outbreaks of 12th Germinal (April 1) 157
And of 1st Prairial (May), 1795 157
Put down and suppressed 158
The power of Jacobinism finally broken 159
Measures of the Government against the Royalists . . 159
Weakness of the State. Tendency to the rule of the sword 159
Great successes of French against the Allies . . . 159
Conquest of Belgium and Holland, September 1794,
January 1795 160
Failure of English descent from Quiberon Bay, July 15-
20, 1795 . . . . 161.

a

PAGE

1795 The Coalition dissolved. Prussia and Spain make peace,
        April, June 1795 161
      Causes of this astonishing success of the Republic . 161
      Continuing weakness of the Republic at home . . 163
      Extreme distress of the great cities 164
      Exhaustion of the Revolutionary spirit . . . 164
      Desire for repose and a settled government. Reaction to-
        wards Monarchy 165
      Constitution of the year III. 165
      The Constitution generally well received . . . 167
      Opposition to the re-election of two-thirds of the Convention 167
      Rising of the reactionary sections of Paris, 13th Vendémiaire
        (October 4, 1795), put down by Bonaparte . . . 168
      The authority of the Convention restored . . . 168
      The military power becomes stronger 169
      Reflections on the course of the Revolution after Thermidor 169

CHAPTER VIII.

THE DIRECTORY. BONAPARTE.

      Character of this period 171
      State of the Republic after Vendémiaire 172
      Policy of the Directory 173
      National Bankruptcy virtually declared 174
1796 Preparations for the campaign of 1796 175
      The campaign of Italy 175
      Bonaparte invades Piedmont from the seaboard . . . 175
      April 28, 1796 176
      May 9-30, 1796 176
      He marches to the Adige. Siege of Mantua. . . . 176
      The Austrians send an army to raise the siege of Mantua,
        and to crush Bonaparte 177
      He defeats Wurmser in a series of engagements . . . 177
      August 3-6, September 1-13, 1796 177
      The Austrians send Alvinzi with a new army . . . 178
      He is defeated at Arcola, November 14-17, 1796 . . 178
1797 Decisive victory of Bonaparte at Rivoli, January 14, 1797 . 178
      Reflections on the campaign. Great skill of Bonaparte . 178
      Character of his strategy 179
      State craft of Bonaparte 180

PAGE

1797 Campaign of 1796 in Germany. Defeats of the French . 180
The Archduke Charles . . . . ., . . . 180
Ability displayed by him 181
Internal state of the Republic; revival of factions . . 182
Royalist and reactionary schemes 183
Coup d'état of the 18th Fructidor (September 4), 1797 . 184
Conclusion of the campaign of Italy 184
Conciliatory policy of Bonaparte; his anti-revolutionary
views 185
He marches on Vienna from Italy 186
Treaty of Campo Formio, October 17, 1797. . . . 186
Gains of France 186
Sacrifice of Venice 187
Reflections on the conduct of Bonaparte 187
Enthusiasm in France at his successes 187
Admiration felt for him in Europe 188
He returns to France, and is received with acclamation . 188
December, 1797 189

CHAPTER IX.

EGYPT AND THE 18TH BRUMAIRE.

1798 The Directory jealous of Bonaparte 190
They engage him to attempt a descent on England . . 191
He proposes to invade Egypt 191
Expedition to Egypt 192
Congress of Rastadt 192
Renewal of causes of discord in Europe 193
Formation of the Ligurian, Helvetian, and Roman Re-
publics 193
Bonaparte lands in Egypt July 1, 1798 194
Battle of the Nile, August 1, 1798, and destruction of the
French fleet 194
Renewal of the war in Europe 195
1799 Murder of the French plenipotentaries at Rastadt, April 28,
1799 195
The Conscription 195
Character of the campaign of 1799 195
Defeats of the French 196

a 2

                                                            PAGE
1799   Formation of the Parthenopæan Republic 196
       Battle of Stochach, March 25, 1799 196
       Battles of the Trebbia and Novi, June 17, 18, and 19, and
           August 15, 1799 197
       The French driven from Italy 197
       Failure of English descent on Holland 197
       Battle of Zurich, September 25-28, 1799 198
       It saves France from invasion 198
       Lamentable internal state of the Republic 198
       Strife of factions 198
       The reverses of 1799 cause all parties to combine against
           the Directory 199
       Weakness and ruin of the State 199
       Desire for a strong Government 200
       Siéyès 200
       Fortunes of Bonaparte in Egypt 200
       He fails at Acre, March to May, 1799 201
       On hearing the news of the state of France, he leaves
           Egypt 201
       Enthusiasm with which he is received on his way to and in
           Paris 202
       He at once becomes the real centre of power . . . 202
       He prepares a *coup d'état* to change the Government . . 202
1799   The 18th Brumaire (November 9), 1799 203
       Scene in Assembly at St. Cloud 204
       Formation of a provisional government. Bonaparte First
           Consul 204
       Character of the Revolution of the 18th Brumaire . . 204
       Reflections on the conduct of Bonaparte, and on the march
           of events 204
       The *coup d'état* was not a crime 205

CHAPTER X.

MARENGO.   LUNEVILLE.   AMIENS.

1800   Wise and healing policy of the First Consul . . . 207
       Financial reforms 207
       Fortunate position of Bonaparte as a mediator between
           factions 209

PAGE

1800   Laws against clergy and émigrés repealed or mitigated . 209
Pacification of La Vendée 210
Rapid recovery of France 210
Constitution of the year VIII. 211
The institutions founded by it 211
Objects of Siéyès 212
Bonaparte First Consul for ten years 212
His Despotism is established, surrounded by merely nominal
restraints 212
Reorganization of the French armies 213
Plans of the First Consul 213
The campaign of 1800 214
Operations of Melas in Italy, and of Moreau in Bavaria . 214
The First Consul crosses the Alps May 16–19, 1800 . . 215
The French army enters Milan, June 2 215
Melas falls back 216
Battle of Marengo, June 14, 1800 217
The French recover Italy 217
Campaign in Germany 217
Advance of Moreau. Ability of Kray 218
Battle of Hohenlinden, December 3, 1800 218
1801   Treaty of Luneville, February 9, 1801 218
Great advantages gained by France 219
Dictatorial tone of Bonaparte 219
1802   Treaty of Amiens, March 27, 1802 221
Great results obtained by the First Consul 221

CHAPTER XI.

THE CONSULATE. RENEWAL OF WAR.

Internal Government of the First Consul 223
The time favourable for reconstructing society in France . 224
Reforms in the State 225
The Judicial system 226
The Code 226
Centralization of local powers 227
Prefects and sub-prefects 227
The Concordat 227
Its effects 229

xxiiCONTENTS.

1802 Public Instruction 230
  General results of these reforms 230
  Changes in the army 230
  Creation of a new aristocracy 231
  The Legion of Honour. Restoration of Titles . . . 231
  Bonaparte made Consul for life 232
  Modification, in a despotic sense, of the Constitution of the
   year VIII. 232
  Partial resemblance of the new Government to that of the
   Monarchy 232
  Its evils 232
  Its merits 233
  Wise Administration of the First Consul 233
  He encourages the movement towards Monarchy . . 234
  Change of manners in France 235
  Foreign Policy of the First Consul 236
  Its craft and ambition 236
  French intervention in Germany 237
1803 Great extension of French power and influence . . . 238
  Disputes with England. March to May, 1803 . . . 238
  Renewal of war with England, May 18, 1803 . . . 239

CHAPTER XII.

THE EMPIRE TO TILSIT.

The First Consul plans a descent on England . . . 241
The flotilla and camp of Boulogne 242
Project of covering the descent by a large fleet in the
 Channel 242
Conspiracy of the *émigrés* against the First Consul . . 243
1804 Execution of the Duke of Enghien, March 21, 1804 . . 244
 This event hastens the movement in favour of Monarchy . 245
 The First Consul proclaimed Emperor of the French, May
  18, 1804 246
 Coronation of Napoleon, December 2, 1804 . . . 246
1805 New coalition against France 248
 Plan of the attack of the Allies 249
 Campaign of 1805 249
 Napoleon quits Boulogne, and surrounds and captures an
  Austrian army at Ulm, October 19, 1805 . . . 249

PAGE

1805 Battle of Trafalgar and destruction of the French and
Spanish fleets, October 21, 1805 250
The project of the descent might have succeeded . . 251
Napoleon marches on Vienna 252
The Grand Army 253
Vienna occupied, November 13, 1805 253
Battle of Austerlitz, December 2, 1805. Ruin of the allied
army 254
Peace of Presburg, December 15, 1805 254
Changes effected by it 255
Austria ceases to be Head of the German Empire . . 255
The Confederation of the Rhine 255
Isolation of Prussia 255
Conduct of that power 255
It declares war against France 256
Campaign of 1806 256
Battles of Jena and Auerstadt, October 14, 1806 . . 256
1807 Ruin of the Prussian army and Monarchy . . ʿ. 257
Napoleon marches against the Russians 257
Winter campaign in Poland 257
Campaign of 1807 257
Indecisive battle of Eylau, February 8, 1807 . . . 258
Peril of Napoleon 258
Reorganization of the Grand Army 258
Decisive victory of the French at Friedland, June 14, 1807 258
Characteristics of these campaigns . . , . . 259
Changes in the art of war 260
Meeting of Alexander and Napoleon on the Niemen, June
ʿ25, 1807 260
Treaty of Tilsit, July 7 and 9, 1807 260
Alliance between France and Russia, and dismemberment of
Prussia 261
Objects of Napoleon in making the treaty . . . 261
His power at its height 262
Extent of the French Empire 262
Vassal kingdoms 262
1807 Allied and subject States 263
The Empire promoted civilisation in some respects . ، 263
Prosperity of France 264
Public works of Napoleon 264
Character of his Government 265

                                                              PAGE
1807   Elements of weakness and decay in the Empire .   .   . 265
       Indignation of conquered nations   266
       Tendency of Germany to unite through common suffering . 266
       Jealousy of Russia.   266
       Decline of the Grand Army in strength   .   .   . 266
       The resources of France unduly strained .   .   . 267
       Moral evils of the rule of Napoleon 268
       Insecurity of his power, which depended mainly on himself 269

                       CHAPTER XIII.

                   THE EMPIRE TO 1813.

       Retrospect of the policy of Napoleon   .   .   . 271
       It changes for the worse after Tilsit 272
       The Continental system, 1807–8   273
       Its mischievous effects upon the Empire .   .   . 274
       It urges Napoleon to make conquests   .   .   . 275
       Project of invading Spain and Portugal   .   .   . 275
       Napoleon dethrones the House of Braganza, November, De-
           cember, 1807   275
       The Royal Family of Spain enticed to Bayonne, and induced,
           to abdicate the throne, May, 1808 276
       General rising in Spain, May, June, 1808 .   .   . 276
       Capitulation of Baylen, July 19, 20, 1808 .   .   . 277
       First appearance of Sir A. Wellesley on the scene. Conven-
           tion of Cintra, August 30, 1808 277
       Great reverses of the French 277
       Napoleon invades Spain and enters Madrid, December 2,
           1808   277
       He leaves the Peninsula at the news that Austria was arming 278
       Campaign of 1809 in Germany 278
       Defeat of the Archduke Charles in Bavaria, April 18, 22,
           1809   278
       Reverse of Napoleon on the Danube at Aspern, May 21, 22,
           1809   279
       Battle of Wagram and victory of French, July 6, 1809   . 279
       Treaty of Vienna, October 14, 1809 279
1810   Napoleon divorces Josephine and marries the Archduchess
           Maria Louisa, March 11, 1810 280
       Successes of Sir Arthur Wellesly in 1809, in Portugal and
           Spain 280

PAGE

1810 His profound insight and military skill   281
Memorable campaign of Torres Vedras, and complete defeat
of the French, June 1810, May 1811   .   .   . 281
Great results of this campaign, and its influence on Europe 282
Agitation in Germany and Holland   282
1811 Murmurs in France   283
Birth of a son to Napoleon, March 20, 1811   .   .   . 283
Jealousy of the Czar, and disputes with Russia   .   .   . 284
Napoleon prepares to invade Russia, November 1811, May
1812   284
Campaign of 1812   284
The Grand Army crosses the Niemen, June 24, 1812   .   . 285
Retreat of the Russians   285
Political mistake of Napoleon in not restoring Poland   . 285
He pursues the Russians   286
Difficulties of the Grand Army   286
Precautions taken by Napoleon   286
He marches into the interior of Russia   .   .   . 287
Battle of Borodino, September 7, 1812   .   .   . 287
The Grand Army enters Moscow, September 15, 1812   . 287
The Russian Governor of Moscow sets fire to the city   . 287
Napoleon delays in the hope of peace   .   .   . 288
Beginning of the retreat from Moscow, October 19, 1812   . 288
Horrors of the retreat   288
Imminent peril of Napoleon and the remains of the army   . 288
Passage of the Beresina, November 25–28, 1812   .   . 289
Napoleon leaves the army for France, December 5, 1812   . 289
Destruction of the Grand Army   289
Reflections on this catastrophe   289
Causes of the ruin of the French   290

CHAPTER XIV.

FALL OF NAPOLEON.

Return of Napoleon to Paris   292
Conspiracy of Malet   292
Defection of York, December 30, 1812   .   .   . 293
Rising of Germany, January–March, 1813   .   . 293
1813 Retreat of the French, February–March, 1813. Energy of
Napoleon   293

PAGE

1813   His immense preparations to repair his fortunes . . 294

Bad condition of the French levies 294

Campaign of 1813 295

Battle of Lützen, May 2, 1813 295

Battle of Baützen, May 20–21, 1813 . ' . . . 295

Success of Napoleon 295

Armistice of Pleistwitz, June 4, 1813 296

Austria proposes terms to Napoleon which he unwisely rejects 296

The successes of Wellington in Spain decide Austria to join the Coalition 297

Battle of Vittoria, June 21, 1813 297

The French driven from Spain 297

Europe in arms against Napoleon 297

His views and objects in the contest 298

Plan of the Allies 298

Battle of Dresden, August 27, 1813 . . . 299

Battle of Culm, August 30, 1813 299

Battles on the Katzbach, at Grossbeeren, and at Dennewitz, August 23 to September 5, 1813 . . ' . . 300

Great battles of Leipsic, October 16 and 18, 1813 . . 300

The French driven to the Rhine 301

Defeats of the French in Italy. Wellington invades France 301

Revolt of the Allied and subject states . . . 302

Desperate condition of the Empire 302

Napoleon thinks only of a death-struggle . . . 303

His preparations 303

The Allies invade France, December 20–26 . . 303

The military situation of Napoleon seems hopeless . . 304

Prostration of France 304

Campaign of 1814 305

Battles of Brienne and La Rothière, January 29 and February 1, 1814 305

Napoleon interposes between the Allies . . . 305

Battles of Champaubert, Montmirail, Vauchamps, and Nangis, February 10–18, 1814 306

Astonishing success of Napoleon 306

Success of the Allies on other parts of the theatre . . 306

Fresh forces raised against Napoleon . . . 307

Battle of Laon, March 9 and 10, 1814 . . . 307

PAGE

1814 Napoleon falls back on Lorraine to rally his garrisons, and
strike the rear of the Allies 308
The Allies march on Paris, March 25, 1814 . . . 308
State of opinion in the capital 309
Capitulation of Paris, March 30, 1814 309
Napoleon dethroned. The Bourbons restored . . . 310
Napoleon hastily retraces his steps 310
He abdicates April 4, 1814 310
Character of Napoleon 311
Reflections on his fall 313

CHAPTER XV.

THE HUNDRED DAYS AND WATERLOO.

1814 Peace of Paris, May 30, 1814 317
Congress of Vienna, September 1814, March 1815 . . 317
Unpopularity of the Government of Louis XVIII. . . 317
Conduct of the *émigrés* 317
1815 Napoleon leaves Elba February 26, 1815 318
He lands in France March 1, 1815 319
His triumphant march to Paris 319
Pacific overtures of Napoleon 320
The Allied Powers declare war March 25, 1815 . . . 320
Great efforts of Napoleon to restore the French army . . 321
Campaign of 1815 321
Two plans of operations open to Napoleon 321
He resolves to attack Blücher and Wellington in Belgium . 321
Concentration of the French army on the frontier . . 322
It advances on June 15, 1815 323
Battle of Ligny, June 16, 1815 323
Battle of Quatre Bras, June 16, 1815 323
Result of the operations of June 16 324
Blücher rallies the Prussians, and moves to join Wellington
on a second line 325
Movements of Napoleon and Wellington on June 17, 1815 . 325
Miscalculations of Napoleon 325
Results of the operations of June 17 326
Great battle of Waterloo, June 18, 1815 326
Defeat and rout of the French army 328
Reflections on the campaign 328
Conclusion 330

LIST OF MAPS.

EUROPE IN 1789. . . . . . *to face page* 23

EUROPE IN 1812. . . . . . „ „ 283

FRENCH REVOLUTION.

CHAPTER I.

STATE OF FRANCE BEFORE THE REVOLUTION.

THE FRENCH REVOLUTION marks the beginning of a new era in the History of the World. A rising in one of the great States of Europe against a long-settled order of things, it overthrew society in France, and wrought violent changes in the Continent; and, at last, directed by military genius, it culminated in domination and conquest, followed ultimately by a terrible retribution. During the progress of this wonderful movement ancient landmarks of reason, of thought, and of faith, were suddenly set aside or effaced; the birth of a new age was ushered in by atrocious deeds of disorder and blood; and in the gigantic strife which ensued the boundaries of Empires were wildly shifted and war was seen in unparalleled grandeur. We are, perhaps even now, too near these events to pronounce with confidence a judgment upon them; yet some of the

General character and results of the French Revolution.

results may be rapidly glanced at. The Revolution has destroyed a great deal that was worthless and in decay in France; it has stimulated the industry and promoted the material progress and wealth of the nation; and it has given better institutions to a large part of the Continent, and removed a number of ancient abuses. Yet, it may be questioned whether, as regards the permanent interests of mankind, this period of confusion, and the rule of the sword, has not led to as much evil as good. At present it seems impossible to form anything like an enduring government in France; faith and loyalty have lost their former power in the land of Coligny, Bayard, and Turenne. Europe, in Napoleon's remarkable phrase, appears destined, from the Tagus to the Volga, to become half Republican and half Cossack. Though wild theories of freedom disturb society in vast tracts of the Continent, true liberty and order have not been reconciled, and Despotism and Democracy are at sullen feud; whole nations have been turned into armed camps, preparing for an internecine struggle; at no period have international rights and the claims and privileges of weak States been so openly held in little respect; and these ominous phenomena may, in different degrees, be all ascribed to the great convulsion which shook the world from 1789 to 1815.

It exhibited vividly the the peculiar   In the peculiar features of the French Revolution we trace plainly the characteristics of the remarkable

people in which it had its origin. No other com- features of the French national character.
munity in Europe, perhaps, would, after a protracted
period of torpor, have so suddenly awakened to agitated life, or so hastily rushed along the path of innovation. In no other community would attempts at
reform have been marked by such rash extravagance
united to many generous aspirations; in none would
theories of Government and Law have been carried
out with equal recklessness and so grave a contempt
of existing facts, and yet have been presented to the
world in such alluring colours. Hardly any other
European nation would have exhibited such vehement outbursts of passion ; would, in all that relates
to political life, have passed so rapidly to opposite
extremes, and oscillated with such uncertain quickness; would with such apparent readiness have
cowered under a ferocious tyranny of which the
crimes cast a deep stain on the French name ; would
have welcomed with such general acclaim a despotism of the sword as the best mode of government ;
would so eagerly have given up the brilliant visions
of a few years before, to follow the phantom of military glory ; or would so carelessly have abandoned
its idol when it seemed to have lost its magical influence. Yet, on the other hand, few communities
indeed, have displayed the noble though unreflecting
ardour seen occasionally in the movement of 1789 ;
have in defending the natal soil against apparently
irresistible odds, given proof of the energy of

1793-4, overrated as that energy has been; have inscribed on their annals such a roll of victories as Rivoli, Arcola, Jena, Austerlitz, Hohenlinden, Friedland, and a hundred more; have made efforts that will compare with those made by France from 1792 to 1815. It should be remembered, moreover, that for evil or good, all these manifestations of French nature were largely due to circumstances of an extraordinary kind; and, probably, but for influences alien to it, the Revolution would have run a less terrible course. In one particular France was true to her general history during this period. Her influence over adjoining countries has at all times been distinct and immense; and it never was so great as when it swept away thrones, dominations, princedoms, and powers, in a pretended crusade for the Rights of Man, and placed the Continent under the feet of Napoleon.

The Monarchy of France before the Revolution; abuses in the system of Government.

But though the qualities of Frenchmen mark the Revolution throughout its progress, we must not suppose that a great change was not inevitable before that event. The institutions of France had ceased long previously to be in accord with the wants of the nation, and the frame of society seemed out of joint, and falling into decay and weakness.[1]   The

---

[1] It has been obviously impossible, in a sketch like this, to comment at length on the state of France before the Revolution, or to describe in detail the institutions of the Bourbon Monarchy, and their working. The number of valuable works on these subjects is so great, that it is difficult to make a selection for the reader. Per-

Government was an ancient Despotism, which gave no scope to political life, or guarantee for rational freedom, and under which the mass of the people were considered as serfs to be ruled at pleasure. Here and there, indeed, the shadows survived of Estates which had controlled the Monarchy, and the Parliament of Paris still retained the semblance of an august authority; but practically, in the greater part of France, the only checks on the will of the Sovereign were either inadequate or simply vexatious. In most of the Provinces the edicts of the King, however oppressive, had the force of law; the Crown generally could impose taxes, imprison the subject without hope of redress, and interfere with the course of justice; and its powers were asserted and carried home everywhere by a system of administration

haps the best general account of the political condition of old France will be found in M. de Tocqueville's *L'Ancien Régime et la Revolution*—see the translation by Henry Reeve, D.C.L., edition of 1873; though I venture to think the picture of abuses and evils somewhat too lightly coloured. Professor Von Sybel's *History of the French Revolution*, though as a narrative dull and one-sided, contains also a valuable chapter on the France of Louis XV. and XVI.; and the whole subject is ably treated in the sixteenth and last volume of M. Henri Martin's *Histoire de France.* The political and social life of the time has been painted with extraordinary force by Mr. Carlyle in his well-known work; and with less justice but with great skill by MM. Michelet and Louis Blanc in their Histories of the French Revolution. Arthur Young's *Travels in France in* 1787-9 contain much valuable information on the economic state of the country; and the *résumé* of the Cahiers or instructions of the deputies to the States-General of 1789 bring out in clear and full relief the innumerable grievances of the nation.

highly centralised, and not seldom corrupt and iniquitous.   Under this scheme of arbitrary rule true
national liberty could not grow up; securities for
public and private rights, long enjoyed in England,
had no existence; and acts of violence and of cruel
wrong were by no means of uncommon occurrence.
It would, in truth, be easy to show that the Monarchy
supplied, in many respects, too faithful precedents to
the Reign of Terror; and wholesale massacres, ruthless deportations, robbery in the shape of forced
contributions, arrests, and detentions without trial,
confiscations, and frauds on the debts of the State,
were known in France before 1793, though Jacobinism
exaggerated in a few months misdeeds previously
spread over centuries.   Nor had the Monarchy, since
the time of Louis XIV., exhibited that regard for
the common weal, and that munificence in the
national interests, which have so often veiled the
evils of despotism in the eyes of the unthinking millions, and are sometimes of real benefit to them.
More than one of the noble creations of Colbert[1] had
been allowed to fall into decay; the great highways
and canals of France were in several districts in a
state of ruin; and, while gorgeous luxury revelled

---

[1] Jean Baptiste Colbert, one of the greatest ministers of Louis XIV.,
born 1619, died 1683.   In addition to the extraordinary impulse he
gave to commerce and manufactures in France, he founded the
dockyards of Brest, Teulon, and Rochefort, almost created the
French navy, and constructed many of the great roads and canals
of the country.

in the palace, the public service was starved and
neglected. Louis XV. probably spent more..money
on his harem than on any department of the State,
and even in his successor's reign it was thought a
marvel that during a struggle with England the
king should have devoted to the fleet a mere fraction
of his princely revenue.

Beside the Monarchy, yet not giving the throne
solid and useful support, were two great Orders
which, at one time, had been supreme in France,
and still held a high place in the State. The Church
raised its front in feudal magnificence, enriched
lavishly with the wealth of centuries, and possessing
immense estates and a large jurisdiction; and its
dignitaries formed a kind of nobility, drawn generally from the great Houses of France, and still invested with many privileges. These patrician prelates and lordly abbots were marked off by a broad
line of distinction from the body of the inferior
clergy; and their haughty demeanour and preten-
tious pride made them generally objects of fear and
dislike. Beside them, spread over the whole of
France, was the numerous Order of the lay Nobles,
who formed one of the most unpopular and worthless
castes that had grown out of the decay of the Middle
Ages. The French Seigneurie of this period still
possessed many of the most odious privileges of Feudalism in the sixteenth century; they were largely
exempted from State Taxation, and enjoyed rights

The Church and the Nobility. Why these orders had become ge- nerally dis- liked.

of the most oppressive kind over their vassals' property and even persons; and yet they were not linked
by the more kindly ties of feudalism to their humble
dependents; for, with many and honourable exceptions, they very seldom lived on their lands, and
squandered their rents at Versailles or in Paris.
They were, too, to a great extent, composed of new
men and needy adventurers, who would have thrown
discredit on any class; and they boasted but few
historic names, illustrious for their services to the
State, and justifying by their past or recent achievements the rank and position of the whole Order.
Can we wonder that such a body as this, at once
tyrannical and ignoble, was viewed in France with
general dislike, and that the eighty thousand families
of which it was composed, which held in thraldom,
perhaps, two-thirds of the peasantry and the soil of
the country, and locust-like preyed upon its resources, were in most instances dreaded and abhorred?
The cause, however, has yet to be noted which perhaps contributed most to expose the Seigneurie to
universal odium.   It was easy enough to become
a noble in France by honourable or dishonourable
means; but the Nobles stood apart, as a class divided
from all below them by the harshest distinctions;
and the result was that they displayed an arrogance,
an insolence, and a contempt of others, not readily
understood in our time.   It was quite usual for the
young noblesse of that day to run down the canaille

of the streets, and to insult the wives of the bour-
geois to their husbands' faces; and, not fifty years
previously, a distinguished seigneur had made it a
grievance that Louis XV. should have administered
to him a mild rebuke for following the pastime of
shooting peasants. Undoubtedly many of the nobility of France were men of a very different kind;
but in the large majority exclusive privilege had
developed its ordinary evil consequences.

Yet, however august the Monarchy seemed, and
high the state of the Church and the Nobles, their
powers, nevertheless, were weak and divided. The
Sovereign of France had not full control over the
ultimate support of all authority, for the Army was,
to a considerable extent, in the hands of princes
and great seigneurs independent of the Crown in
different degrees; and the soldiery, shut out from
promotion and reward, and subjected to a cruel and
degrading discipline, had long been filled with elements of disaffection. The power of the Monarchy,
too, was crossed and thwarted by the decaying remains of feudal institutions no longer capable of
doing good; and it had lost a great share of the
patronage of the State, and of the administration of
the public service, through the ruinous practice of
selling offices which had gone on for several genera-
tions. Thus while the prerogatives of the King
were immense, and were often exercised with extreme harshness, the Monarchy was deficient in

The power of the ruling orders was divided and decayed, and consequently weak.

essential strength; its action was impeded in many
spheres in which its influence should have been
absolute; and it was, in a great and dangerous degree, deprived of the right of an Executive Government to select and dispose of its own instruments.
As for the Church, virtue had gone out of it, and it
was enfeebled by imbecility and discord. The
hierarchy might boast of their sounding titles, and
walk in purple and fine linen, but their moral
influence had become nought; and though they
could still torture oppressed Huguenots, and condemn heretical books to the flames, they were unable
to stem the tide of thought that was sweeping away
their proud pretensions. Besides, little sympathy
existed between these potentates and the lower
clergy, divided from them by a broad barrier; and
while in the high places of the Church no Bossuet
stepped out to do battle with Voltaire, in many
dioceses the village curés were secret enemies of
their superiors, and hated the ecclesiastical system
around them. The condition of the nobility, too,
was one of weakness and internal dissension. In a
State ruled as France long had been, the Order had
little political power; and though it possessed most
unjust privileges, and extravagant and oppressive
local rights, its authority was exceedingly small in
all that related to the central Government. The
Nobles were not a great aristocracy with a potent
voice in the national councils; they had no part in

the work of legislation; and their influence was scanty and jealously curtailed in many departments of the public service, and even in the administration of the country districts. A tacit feud, too, existed between the more distinguished and inferior nobility: the Montmorencies and La Tremouilles despised the crowd of new and petty seigneurs whose pretensions seemed a disgrace to their own ; and enlightened and liberal members of the Order condemned the insolence, tyranny, and greed, of the great mass of the men who called them fellows.

It should be observed, moreover, that the different Orders which embodied power and grandeur in France had been repeatedly in angry collision, and their representatives had largely incurred discredit. A quarrel between the Crown and the Nobles had come down from the days of Louis XIV., and had raged at intervals during the reign of his successors. The Parliament of Paris, too, had, more than once, risen against royal assumptions or claims; and Louis XV., at the bidding of a harlot, had treated this body with odious severity. The Church, besides, had had many squabbles with what may be called the party of the Court nobles, and the Parliaments of the kingdom had often resisted its bigoted and intolerable assumptions, though they had joined zealously in Huguenot persecutions, and, with the characteristic feelings of lawyers, had steadily refused to make changes in a barbarous

Dissensions between the Crown, the Church, and the Nobles, and discredit of all authority.

scheme of criminal procedure. It is easy to estimate
the effects of these open and public conflicts; the
authorities of the State laid bare the weakness and
vices of the institutions of France to the eyes of a
nation deprived of its rights, and the growing contempt that was felt for them, and their unpopularity
largely increased. As to the reputation of the
higher Orders in France during the greater part of
the eighteenth century, it is almost superfluous to
refer to it. The orgies of the Regency can be only
compared with those of the worst pagan Cæsars; and
Louis XV. was a degraded being, a slave of coarse
and unmanly vices, and a puppet of scheming priests
and mistresses, all the more despicable because not
their dupe. As for the nobility, whether in Church
or State, some, doubtless, were blameless and illustrious men, but the great majority were only conspicuous for dissoluteness, extravagance, and frivolous
luxury. The scandalous and not concealed de-
baucheries of cardinals and bishops had been common talk; and the ordinary life of the better of the
class was a graceful round of refined amusement, of
idleness, and of epicurean indulgence. Decline,
however, was most apparent and beyond remedy in
the lay nobility. Even among the historic families
of France hardly a name of real eminence appeared;
and the Richelieus and Condés, who had built up
the realm, had degenerated into fops and courtiers.
As for the great mass of the Nobles, they were dif-

ferent beings from the chivalry of Rocroi and Landen;
very few had won honour in the field; their manhood
was wasted in gambling and intrigues, in mere display and effeminate pursuits; and their ignorance
and listlessness were on a par with their overweening
arrogance and conceit. Not many of the young
gallants of this time could write even a common
letter; and pulling out the threads of silk tissue
seemed to fine ladies the business of a day. Contrast in this, and in all other respects, the characteristics of the aristocracy of England trained under
the discipline of public life, and taught to discharge
high social duties by a vigilant and exacting general
opinion.

Under this array of grandeur and state, iniquitous and oppressive, but really weak; with splendid
traditions, but no vital strength; with lofty pretensions, but failing and decried, was marshalled the
ill-governed Nation : the twenty-five millions of the
French Commons, who, it was said, 'counted as
nothing in France.' Discordant elements, however,
lurked in the mass; and it is necessary to perceive
this truth, or we shall never understand the events
that ensued. A great Middle Class had grown up in
France, especially in the principal towns; and the
professional and mercantile Orders, reaping the fruits
of ages of honourable toil, and not spoiled by luxu-
rious idleness, had, in numberless instances, amassed
wealth, and attained to a high degree of refinement.

State of the
Commons
of France.
The middle
classes; how
they were
cut off from
the People,
and the re-
sults.

An aristocracy of riches and culture had been formed
among the lawyers, the physicians, the manufacturers, and other traders; and though it had been
trained in a bad school of thought, and was wholly
wanting in political knowledge, it really comprised
what was most enlightened, most intelligent, and
truly sound in the kingdom. This numerous and
respectable class disapproved of the existing order of
things, in which they were esteemed inferior beings,
and especially regarded the Nobles with dislike,
whose insolence was often directed against them.
But, as regards the mass of the Nation, they too
were isolated and stood apart; and a whole series of
invidious distinctions cut them off from the People
from which they sprung. In fact, Feudalism had
left its stamp on this as on other Orders of the State,
and a policy of distinction had made the mark
deeper. The professions, the trades, and the indus-
tries of France, were organised on a system of exclusive privilege, and of grasping and close monopoly;
and they formed everywhere a number of Corporations, of guilds, and of separate castes, with distinctive rights and peculiar immunities. The Bour-
geoisie, as they were contemptuously called, were
thus led to look down with scorn on the bulk of the
community around; and though eminent men were
among their ranks, they had little sympathy with
the great body of the People, and they were viewed,
as a general rule, with envy and ill-feeling by their

poorer dependents. One of the cardinal facts of this period is the alienation that existed between the great employers of labour in France and the workmen and artisans of the towns, and it marked many phases of the Revolution.

In France, therefore, deep lines of distinction divided the Upper and Middle Orders, and separated both alike from the People. In no country were differences of class more offensively marked, and yet more unpopular; and Feudalism, which had knit society, in one of its stages, in close dependence, now broke it up into disunited fragments. We turn to consider the mass of the Nation, which may be viewed in its two chief parts, the inhabitants of the country, and those of the towns. The occupiers of the soil, as a general rule, were subjected to heavy and vexatious taxation, from which the owners, if noble, were exempt; and, in the distribution of the imposts of the State, were frequently treated with harsh injustice.. Yet these grievances were small compared to the burdens their lords imposed upon them, and to the usage they had often to endure. In a very considerable part of France the peasantry held the land by a permanent right; Condition of the Peasantry. and in some of the Provinces, especially in the North, large farms were cultivated under long leases. Though the peasant estates were greatly subdivided, these districts were comparatively thriving; and they were easily distinguished by the signs of com-

fort and of agricultural progress evident in them.
Yet the cultivators even of these favoured regions
were liable to numberless petty vexations, to iniquitous restrictions upon their industry, to services
sometimes degrading and mean; and their very
prosperity caused them to resent the imperious
harshness and neglect of duty of their generally
absentee superiors.   In other parts of France the
land was held, over most of its breadth, by precarious tenures; and, except in a few Provinces,
where the relations between the lord and vassal were
still not unkindly, the occupiers were a race of serfs,
ground down by rack-rents and feudal oppression,
kept in abject dependence, exposed to wrong, and
often struggling' on the verge of destitution.   The
districts held on these wretched conditions seemed,
in spots, smitten as it were with barrenness; and an
experienced eye-witness[1] wrote that, except in Ireland—at that time in her very worst state—he had
never beheld such squalor and misery.   In these
tracts the food of the peasant was not seldom nettles
and pulse; and it was in these that, a few years
afterwards, rose the troops of half-clad and ferocious
savages, who, at the first whisper that deliverance
was near, burned the châteaux of the abhorred seigneurs, on whom they laid the charge of their sufferings.   The condition of the lowest tillers of the soil
must have been pitiable in this state of things; and

[1] Arthur Young.

there is reason to believe that the wages of the agricultural labourer in France before the Revolution were not half (in some places) what they have since become. As for the population of the towns, it had been allowed to multiply densely in ignorance and want; and the large cities swarmed with dangerous classes—poor, discontented, and enemies of the rich, who stood selfishly aloof from them—and blindly eager for any change. It was from these orders that Jacobinism recruited its armies of devastation and crime—the murderers who crowded the prisons with corpses—the furies who shrieked round the guillotine.

of the population of the towns.

It should be remembered, too, that while the whole frame of Society in France was thus ill-ordered, and its component parts were weakened and diseased, a spirit of wild innovation had grown up, which fiercely assailed the tottering structure. Thought, at least among the privileged classes, had long been comparatively free; and it had embodied itself in a brilliant literature, audacious, sceptical, yet unreflecting, which held up almost every institution, and even the existing order of things, to universal contempt and ridicule. The movement, doubtless, originating in the inexperience in political life of French men of letters, like all other Frenchmen, received a definite character from two men of genius; and while Voltaire sapped away the throne, the altar, and the privileges of the great by keen

Destructive and sceptical tone of contemporary French Literature.

c

satire and malignant wit, the more profound Rousseau constructed theories for the regeneration and
happiness of mankind which bade defiance to all
social arrangements. Thus intellect, which, in a
healthy State, is always upon the side of order, and
aims only at temperate reforms, became destructive
and anarchic in France ; and though it would be a
mistake to think that it caused the Revolution, the
causes of which lay much deeper, it accelerated that
event, and left its mark upon it. Strange to say,
too, so little were the signs of the coming time
understood in France, this godless, false, and spurious
philosophy found high favour among the classes
destined to suffer most cruelly from it. It had become the custom before the Revolution to scoff at
faith, to decry the past, to relish attacks on people
in high places, to complain of the absurdities of class
distinctions, to see in the complication of old laws
and customs a mass of rubbish to be swept away, to
put together pretty and ingenious systems for making
the world a scene of pleasure ; and lettered mar-
quises and brocaded dames babbled of the philosophic
dictionary, and the *contrat social,* as though doctrines fatal to their pretensions at least were the
very perfection of truth and wisdom. Of all the
phenomena of this period, none is more instructive
than this curious spectacle of the natural supporters
of social order playing with the instruments that
were to hasten its ruin.

This sketch of France before the Revolution may The progress of evil continued down to the Revolution.
seem overcharged with dark colours as respects the
immediately preceding period. But the sympathy
which the appalling fate of thousands of unhappy
victims evokes ought not to blind us to the fact,
that the worst evils of which we have given a brief
account were never more apparent and active than
during the reign of Louis XVI. The King was
certainly well-meaning; but the Monarchy was
seldom guilty of acts more arbitrary, violent, and
iniquitous than those sanctioned by Brienne [1] and
Calonne.[2] At no time were the imbecility of the
State and the dissensions between the privileged
Orders more plainly evident than when Louis was
twice compelled to dismiss his minister, at the
bidding of an insolent party of the Noblesse, and
when the Sovereign, the Notables, and the Parliament of Paris, were in angry collision on such subjects as the Public Debt and national bankruptcy.
At no time, too, was the throne more weak than
when Princes of the Blood were conspiring against
it, or when the unjust reforms of St. Germain[3] had

[1] Loménie de Brienne, Archbishop of Toulouse and Sens, born 1727, died 1794. As minister of Louis XVI. he exiled the Parliament of Paris to Troyes in 1787, and compelled it to register edicts of the King which it had opposed.

[2] Charles Alexandre Calonne, Minister of Louis XVI., born 1734, died 1802. He wasted recklessly the finances of the State, quarrelled with the Notables whom he had convened in 1786, and is believed to have promoted the policy afterwards followed.

[3] Claude Louis St. Germain, born 1707, died 1778. As minister of war to Louis XVI. he introduced a harsh and degrading discipline

c 2

increased the growing discontent of the Army. Nor
at any period were the high ecclesiastics and Nobles
of France more widely unpopular than when they
clamoured against Necker and Turgot,[1] and, at a
season of national distress, refused to submit to
equal taxation; nor had the lives of the class
mended, though vice was, doubtless, less gross and
cynical than in the days of Louis XV., and the arrogance of the great was sometimes tempered by a
condescension not less insolent. Never, too, were
the distinctions of class more sharply defined than
at this juncture, and felt with more bitter resentment; and at no time had the poorer classes, in
consequence of numerous bad harvests, suffered more
hardships or been worse treated. Nor was the spirit
even of the central Government patriotic, or in the
general interest: its feeble attempts at superficial
reforms were, for the most part, purely selfish expedients; and, even when judicious, they were abandoned at the first symptom of class opposition. The
King and Queen, too, did not escape the discredit

in the French army; and by his advice commissions were afterwards more restricted than they had ever been before to the class of
the nobles.

[1] For an account of the policy of Necker and Turgot, and of their
plans of reform, see M. Henri Martin's *Histoire de France*, vol. xvi.
Speaking generally, the measures they advocated were the abolition
of the monopolies that sapped French industry, an equitable distribution of taxes instead of the injustice that prevailed, retrenchment,
the publication of the finances of the State, decentralisation, and
something like representative assemblies. Turgot, however, was
a much abler man than Necker.

which had fallen on all dignities; the weak character and awkward demeanour of Louis XVI. provoked contempt; and though the life of Marie
Antoinette was pure, foul scandal had gathered
around her name.   The period, it is unnecessary to
say, was especially one of wild speculations and of
shallow schemes of universal change; and, in a
word, all the elements of ill which had been gradually
collecting in France had lost none of their fatal
power.

Such, then, briefly was the condition of France
before the crisis of 1789.   English statesmen[1]
trained in political life had long before seen that a
change was at hand ; but no minister of Louis XVI.—
few, as far as we know, among thinkers in France—
had the least apprehension of coming danger.   Many
in high places, as we have seen, were intent on
sweeping abuses away, and had planned magnificent
schemes of reform ; and the announcement that the
States-General would meet seemed the dawn of a
new and golden age to thousands foredoomed to
death or exile.   Even the agitation which followed
that event—the disturbances which broke out in
several provinces, the blind stirrings of the unen-
franchised millions, dull, unintelligible, and yet
ominous, and the clamorous exultation of the middle

Brilliant anticipations of the future in France.

[1] As far back as 1753 Lord Chesterfield had written : ' All the
symptoms which I have ever met with in history, previous to great
changes and revolutions in government, now exist and daily increase
in France.'

classes, who thought their time of hope had arrived—
did not dissipate these illusions; for it was believed
that the genius of a polite age would not allow
popular excesses or passions. With these shallow
and idle visions—the growth of ignorance, and of
the false sentiment which pervaded a society in an
unhealthy state—was mingled much that was truly
noble, many generous and high aspirations; and a
glowing rainbow of deceitful hope shone brilliantly
over the dark torrent that was carrying old France
to the depths below.

CHAPTER II.

THE STATES-GENERAL AND NATIONAL ASSEMBLY.

ON May 5, 1789, the States-General met for the first time at Versailles. More than a hundred and seventy years had passed since, in the youth of the Bourbon Monarchy, this ancient Assembly of the Estates of the Realm had consulted upon the common weal; and they were now convened for the same purpose when that Monarchy was in decline and peril. The spectacle formed an imposing sight, and it seemed for a moment as if the elements of the long discordant community of France had blended in happy and auspicious union through the representatives of its different Orders. A great hall had been laid out in the palace, and prepared in stately and magnificent pomp; and royalty welcomed the National Estates, composed of more than twelve hundred deputies, with a splendour worthy of the solemn occasion. The King, with the ministers of State, in front, and the Queen and Princes of the blood at his side, sate on a throne brilliant with purple and gold; below, arrayed in separate processions, spread the ranks of the Nobles, all plumes and lace; of the Commons, in homely and

Meeting of the States-General.

simple garb; of the Clergy, the superb robes of the
prelates mingling strangely with the cassocks of the
village priests; and from galleries above a throng of
courtiers, of jewelled dames, and of foreign envoys,
contemplated curiously the interesting scene.   Outside, crowds of eager spectators filled the balconies
and covered the roofs of Versailles, decked out gaily
for a brilliant holiday; and the groups extended as
far as the capital, already stirring with passionate
excitement.   All seemed deference, goodwill, and
hope, when the King announced that he had called
together the wisdom of France to assist at his counsels; and even a declaration that his chief object was
to provide for the pressing wants of the State did
not weaken the prevailing sentiment.   Yet it
was observed with regret that the face of the Queen
seemed overclouded with settled care; and jealousy
had been aroused in more than one breast by the
distinctions drawn by the officials of the Court, and
by the contrast between the feudal magnificence of
the nobility and the lordly hierachy, and the plebeian
aspect of the meanly-attired Commons.

The Commons declare themselves the National Assembly.   On the following day the Estates were invited,
their first sitting having been merely formal, to
meet again for the despatch of business.   The intention of Necker, the chief minister, had been to
convene them for the object mainly of procuring
supplies for an exhausted treasury—an increasing
deficit had for many years been one symptom of the

ills of the State—but it had long been arranged
that they were to advise on the administration and
general affairs of the kingdom. A preliminary
question, however, arose, which brought out distinctly the deep-seated differences already existing
in this Assembly. According to ancient precedent,
the separate Orders of the States-General gave their
votes apart; and the Nobles and Clergy, if they
coalesced, could easily neutralise the will of the Commons, voting being by orders and not by persons,
and the votes of two orders being thus decisive.
Trusting to this usage, the Court had consented, in
the elections which had lately taken place, that the
number of the representatives of the people should be
double what it had been formerly; for it was thought
no danger could possibly arise, and the concession
was a popular measure. The Commons, however,
had made up their minds not to be reduced to ciphers
by ancient forms; and they insisted, accordingly,
that the three Orders should hold their delibera-
tions apart, and that votes should be given by head;
that is, be determined by the majority of individuals
in the collective Assembly. The Nobles protested
against this scheme, being but three against more
than six hundred Commons; and they resisted an
invitation to a fusion in which their influence might
be diminished, the three hundred clergy, though
divided in mind, siding with them at the command
of the bishops. During several weeks the separate

Orders stood sullenly aloof and almost hostile, and nothing in the nature of business was done, to the mortification of a minister and a Court exceedingly in need of a supply of money. The Commons, however, held firm, backed by messages from the provinces, and by the attitude of the great neighbouring city, already effervescing with agitation; and at last they adopted a decided course. On June 17, it being known that some Liberal nobles were on their side, and several of the inferior clergy having come to them, they declared themselves the National Assembly of France; and, while they invited their fellow members to join them, announced that nothing should prevent their proceeding ' to begin the work of national regeneration.'

Three days after this important event, the Commons found, to their extreme surprise, the great hall at Versailles, in which they had sate, shut up; and the Grand Master of the Ceremonies curtly told Bailly—a distinguished member whom they had chosen president—that the place was wanted for the royal convenience. Alarm was seen in many faces, for a sudden act of violence was feared; but, at the instance of one or two courageous men, the whole body betook itself to an old Tennis Court at a short distance, and, amidst a scene of passionate excitement, swore a solemn oath ' that it would never separate until it had set the constitution on a sure foundation.' (June 20.) Meanwhile the Court

The oath of the Tennis Court, June 20.

had been forming schemes for dealing with these extraordinary proceedings, and for putting an end to a state of things which appeared to it the wildest presumption. Necker, timid and cautious, proposed a compromise, to which it is said the King inclined; but the counsels of an extreme party prevailed, and it was resolved to make a display of vigour. On June 23, having been kept standing by official insolence for some time under rain, the Commons were summoned again to the great hall; and the King read them a lecture, which had been put into his mouth, to the effect that it was his pleasure that the three Orders should, as in old times, deliberate and vote apart, and that, if further resistance were made, ' he would do by himself alone what was meet for his people.' This foolish harangue was met in silence; but when the Grand Master of the Ceremonies, following, it is said, the etiquette of the ancient depotism, commanded the Assembly to depart, he was told by Mirabeau [1]—a man whose

---

[1] Honoré Gabriel Riquetti, Count of Mirabeau, born in 1749, was the grandest and most striking figure of the first part of the Revolution. This extraordinary man was a noble by birth, but, like many other French nobles, had joined the party of innovation. This attitude was in part caused by the antecedents of a career of vice and recklessless, marked, however, by evidences of real genius, in which he had quarrelled with his family, and been persecuted at Court. His powers as an orator were commanding; and though he stooped to become a demagogue, he had true political sagacity and insight, and many of the highest qualities of a statesman. Many of the most serious charges of contemporaries against him seem to be without foundation.

pen and voice were already a power in France—that ' they were met there by the will of the people, and that bayonets alone should drive them from the spot.' In a few moments a vote was passed by acclamation, which declared the persons of members of the Assembly sacred, and made it a capital crime to molest them.

These bold measures, supported as they were by popular demonstrations in Paris, intimidated the Court, which thought that the Commons would be silenced with as much ease as the old Parliaments had been by Beds of Justice, the coups d'état of the Bourbon Monarchy, by which the Sovereign had often put down opposition in these feudal Assemblies. The King, when apprised of what had taken place, remarked, it is said, only, ' Let them stay if they please.' With his usual weakness, he allowed himself to float passively on the tide of events. Before this time a considerable number of the minor clergy had joined the Commons; and they were soon followed by the party in the Nobles which wished for reform, and even longed for change. The rest of the Order, however, still held aloof; but at last, at the request of Louis himself, they gave up an opposition that was becoming fruitless, and fell into the ranks of what had now been fully recognised as the National Assembly. This step, however, had been taken in order mainly to conceal arrangements by which the extreme Court party thought they would triumph and overawe the Commons they feared, yet despised. On July 11, Necker, whose advice to convene the States-General had made him very popular, whatever his motives were, was dismissed; a ministry of soldiers and of re- actionary nobles, either unknown or disliked, was set up; and the Assembly saw, not without alarm, that batteries were being constructed at Versailles, and heard that troops were approaching in thousands, and that an armed force of irresistible strength was being directed upon the capital. Rumour spread, too, that it had been said in the palace ' that the best place for a mutinous Assembly was a garrison town where it could be kept under,' and that the Queen had shown her children to noble officers, and had asked, ' Could she rely on their swords ? ' and there was a report of what was described as ' an orgie ' in which ladies of honour had done strange things to enthral youthful dragoons and hussars.

Dismission of Necker, and forma- tion of a reactionary Ministry.

This intelligence, magnified by a thousand tongues, quickened the already fiery excitement of Paris, and the flame soon rose into a conflagration. On July 12 proclamation was made ' on the part of the King' to keep the peace; and, presently, soldiery with strange faces—the half-foreign German and Swiss regiments, of which there were several in the royal army—were seen occupying the central streets and chief squares of the great city. The sight caused terror and indignation ; angry meetings were

Rising in Paris.

harangued in the gardens of the Palais Royal by
passionate speakers; and a procession was formed
carrying at its head busts of Necker and of the Duke
of Orleans,[1] whose largesses and opposition to the
Court made him one of the idols of the low populace.
In a charge made to disperse this assemblage, the
Germans cut down one or two men of the French
Guards with a few unarmed persons; and the foreign
uniforms were ere long seen in the avenues of the
Tuileries driving before them a scattering collection
of citizens in flight. These incidents, not in themselves momentous, proved the spark that reached
the combustible mass, and fired it in a wide-spread

Mutiny of the French Guards and insubordination in the Army. explosion. A spirit of disaffection—the natural result of a brutal discipline, and of harsh treatment—
had shown itself in the French Guards, as, indeed,
in other parts of the Army; and as it was very apparent in a body exposed to the allurements and mob

[1] Philippe, Duke of Orleans, born in 1745, was one of the most
infamous personages of the Revolution. This Prince combined in
himself all that was most depraved and bad in the old noblesse, and
all that was most odious in the ambitious mob leaders. Having
become out of favour at Court, in part on account of his personal
cowardice, he revenged himself by circulating slanders against the
Queen, joined the party of reforming nobles, and laid himself out
to gain popularity in Paris by flattering the populace and by a display of extravagance. He became afterwards one of the noisiest of
the Jacobin leaders; and between 1789 and 1791 combined more
than once, for his own selfish ends, against the throne, and even the
life of Louis XVI. His complicity, however, with the crimes of the
Revolution did not atone for his royal birth; and though he
paraded the name of Egalité which he had assumed, he perished
during the tyranny of Robespierre.

speeches of Paris—for the Guards were part of the
city garrison—the men had been lately confined to
barracks.   When the news arrived of the fate of their
comrades, the Guards broke out and fired at the
Germans ; and the first example of military insubordination caused the dissolution of all military
authority.   Shouts of ' Long live the Nation ! ' were
heard from the quarters of regiments usually stationed
in the capital ; .even the foreign troops were affected
by the general contagion in a few hours, and sullenly
declared they would not shed blood ; and the only
resource left to the indignant officers was to withdraw
the demoralised mass, and to beat a retreat.   A thrill
of exultation ran through Paris at the disappearance
of the strange invaders ; and power once dreaded
having proved worthless, disorder and violence were
let loose.   During the night the city was wildly astir ;
the dark swarms of poverty and vice, which became
afterwards the legions of the Reign of Terror,
emerged in thousands from their wretched haunts,
mingled here and there with less hideous groups ;
and shops were sacked, and the great Town Hall invaded by these mobs to the cry of ' Arms ! '   Next
morning a provisional committee, composed of the
chief men of the sixty districts into which Paris had
been divided, took the rule of the capital into their
hands, the old authorities having proved powerless ;
and an endeavour was made to give a kind of organization to the movement, and in some measure to

direct and control it. The citizens were encouraged

The Commune, the National Guard.

to form themselves into a militia of volunteers drawn from the districts; these bands were to wear in their cockades the Parisian colours of blue and red; and they were not only to find arms as best they could, but arms were liberally supplied to them. M. de Flesseles, head of the old Town Council, was made president of this board; and, though the objects of the members varied, a general intention certainly prevailed to keep the insurrection within bounds. Such was the origin of the world-renowned Commune of Paris, and of the National Guard, names of deep significance in the Revolution.

Although partly controlled by these means, the revolutionary movement went on throughout the day with terrible energy. The levies of the district started into life, and were enrolled into the new civic army; the streets bristled with forests of pikes; arms were violently seized wherever they were found; and mobs were seen trailing antique cannon, and tossing about pieces of feudal armour, torn recklessly from arsenals, with swords and muskets. On the whole, however, the better class of citizens predominated in the National Guard, and checked the excesses of the lowest populace; and though it was accelerated by such events, the time had not yet come when the violent elements of society were to overpower all others. The presence of this better order of men in the ranks is strong proof of the general indignation felt at the late demonstration made by the Court; and the rising was anything but the mere work of a mob set on by a few designing leaders. In the afternoon the French Guards, to a man, went over to the popular side, their terrified officers protesting in vain; and, amidst wild shouts of passionate exultation, they were made grenadiers of the National Guard, and played an important part in the events that followed. On the 14th a great crowd entered the courtyard of the Hospital of the Invalides—a noble establishment, like our own Woolwich—and the governor was obliged to allow them to take the vast store of arms laid up in the arsenal, for the inmates passively seconded their efforts. By this time nearly 80,000 men had been marshalled more or less regularly; and as no signs of resistance appeared, they were encouraged to acts of more open daring. On the verge of the quarter of Saint Antoine rose the celebrated fortress of the Bastille; and it was resolved to attack this dreaded place, the very emblem of ancient despotism, and infamous for its mysterious horrors. An armed mass poured down to the spot, and after an ineffectual attempt at a parley, the drawbridge was passed, and the inner court reached, close to the eight frowning towers of the hated dungeon. A discharge of musketry drove the assailants back; but cannon were brought up by the late French Guards, and a white flag before long was waved

Siege and storming of the Bastille, July 14, 1789.

D

from the ramparts, the commandant, Delaunay, having been compelled by the garrison (alarmed or ill-disposed) to surrender. The victors rushed into the ancient den, amazed at the feat they had accomplished, and carrying out many of the arcana of the place—old instruments of torture, and prison records; but their victory was not unstained by cruelty. The greater part, indeed, of the garrison were set free; but Delaunay and several of his men were murdered, and their heads were borne on high on pikes—the first of many subsequent scenes of the kind. De Flesseles, too, was attacked and shot, for a tale spread that he had deceived the people; and several other deeds of blood were committed. As yet, however, the better part of the National Guards maintained comparative order; and the extraordinary and rapid changes which had occurred had rather proved the weakness of the royal authority than brought out anarchy in its most frightful aspect.

Such was the end of this sorry attempt to work a violent change in the State, to intimidate the Assembly, and to overawe Paris. The result had been to hand the capital over to unknown and revolutionary forces, and to prove that no trust could be placed in the Army, the chief and, usually, the sure instrument of power. The extreme Court party stood furious and aghast; and the Count of Artois, the younger brother of the King, and the Charles X. of a later age, with two other magnates of a like stamp,

declared that these things were not to be borne, and
hastened indignantly over the frontier. This was
the beginning of the emigration—that desertion of
the King by his natural supporters which was one of
the many evil features of the time, though the circumstance will not be surprising to those who know
what little genuine sympathy existed between the
nobles and the Crown. Meanwhile the Assembly
had loudly condemned the violent measures attempted
by the Court; and Mirabeau alluded, in no am-
biguous terms, to the part, said to have been taken
by the Queen in a project 'worthy of St. Bartholomew.' The King, shifting in the usual way, hastened
to make peace with the stronger side, dismissed the
ministerial cabal, and recalled Necker; and the Assembly listened with sincere good will to the explanations of an amiable being whose principal
fault was weak simplicity. Soon afterwards a deputation from Paris invited him to pay the city a visit;
and the Monarch assented, though Marie Antoinette,
indignant at the affronts given to royal authority,
and knowing how unpopular she was herself, en-
treated him with tears not to make the attempt.
The citizens, however, proud of their triumph, received their Sovereign with acclamation; and an
hint in an address that he had been 'conquered'
was treated graciously by Louis as a joke. All that
had lately occurred was sanctioned by him; the provisional committee received the name of the Com-

Beginning of the emi-gration of the Nobles.

Recall of Necker.

The King sanctions what had been done in Paris.

mune of Paris, with immense powers; and Bailly, the
president of the Commons, was appointed mayor;
while the young Marquis of Lafayette, one of the
enthusiastic reforming nobles, was appointed commander-in-chief of the National Guard. In sign of
reconciliation, the white colours of the House of
Bourbon were added to the blue and red of the
The Tri-
colour Flag.
capital on the ensigns of this force; and thus
originated the Tricolour Flag, which Lafayette, with
conceit or foresight, exclaimed ' would soon make the
round of Europe.' Though two or three bad instances of violence followed, tranquillity seemed established in Paris for a time; the king returned well
pleased to Versailles; but, between impotent threats
and feeble concessions, how much of the divinity
remained that hedged round the Monarchy?

Risings in
the Pro-
vinces.
Notwithstanding this quiescence, however, the
events which had taken place in Paris went like an
electric shock through the kingdom. The influence
of the capital of France over the provinces has always
been very great, and it acquired additional power at
this juncture. The sudden collapse of the majesty
of the throne, the successful triumph over ancient
authority, and, above all, the revolt of the troops,
stirred the minds of men to their very depths, and
long pent up elements of hate and confusion broke
out in many places in appalling strength. In the
southern, midland, and south-eastern districts, where-
ever Feudalism was most oppressive, wherever misery

was most keen, the peasantry rose against their lords; and from the Rhone to the Loire there was a great blaze of châteaux, the infuriated vassals tossing into the flames the charter-chests and muniments which contained the records of privileges no longer tolerable. A few murders of seigneurs also took place; even in the north the payment of rents and the customary services were generally resisted; and, wretchedness adding force to the movement, bands of squalid savages in some provinces ' descended from the hills, destroying the corn, plundering orchards, and doing all kinds of mischief.' Many of the towns, too, showed signs of insurrection, clamouring for an extension of municipal rights, and for an abolition of old monopolies; and violent bread and meal riots were frequent, for the year was one of peculiar scarcity, and the sufferings of the poorer classes were extreme. Nor was the capital itself free from causes of disturbance and trouble. Order was, indeed, maintained by Lafayette; Bailly, the mayor, laboured to please the citizens by civic pomp and gay exhibitions emblematic of their newly-acquired liberties; and the Commune, now formed into a body of three hundred members, made efforts to supply the wants of the poor, to find employment for artizans out of work, to cope with the difficulty of increasing poverty. But, as always happens on such occasions, the new powers were decried by envious demagogues, the more bitterly because they were new; and of what avail were displays of fireworks, enthusiastic 'festivals of the Bastille,' 'trees of liberty' rising in gardens and avenues—nay, even doles, offerings, and all the expedients of a merely improvised system of relief—to thousands of hungry men and women? Between agitation and the presence of want, Paris was soon fermenting with elements of disorder, all the more dangerous because as yet suppressed.

Legislative measures of the Assembly. These tidings of evil came to interrupt the consultations of the National Assembly. It had been engaged in economic discussions as to the best means of meeting the deficit, and as to framing a new constitution for France; and high-sounding principles of reform, conceived in the spirit of the new philosophy, had been already hailed with applause. The measures it adopted to remove or palliate the stern practical ills it had now to face were, in part, conceived in a generous spirit, but were characteristic of the national temperament, and too plainly revealed the political ignorance and passion for change that widely prevailed. As for the towns, the Commune of Paris was encouraged in doing whatever it pleased; and Bailly and Lafayette were thanked for their well-meant and patriotic efforts. Little was attempted in the case of other towns, except to give 'promises of free trade;' but the middle classes were allowed, or invited, to put down disorders, by themselves, by force; and in this manner an armed organisation of National Guards was spontaneously formed in almost all the great cities of France, self-elected and independent of the State. A great and sudden revolution, however, was effected in the social relations of the whole rural community in the kingdom; and the imposing edifice of antique Feudalism was thrown down in a moment, and laid in the dust. One or two nobles, on the Liberal side, having drawn a frightful picture of feudal abuses, the Assembly, in spite of a few protests, started to its feet and declared, almost to a man, that this state of abominations should cease; and resolutions were passed, in a single night, abolishing claims that had been the growth of centuries, and involving in a common extinction the most barbarous remnants of cruel serfdom, with tithes, quit-rents, and similar dues. The sitting closed with enthusiastic shouts, a Te Deum mingling in strange accompaniment; and though distinctions were afterwards drawn between such privileges as that of the lord bathing his feet, when cold, in the blood of his vassals, and others of a more modern kind, an opposition formed by the Nobles was overborne by an increasing majority, the Commons and lower clergy ruling the Asssembly; and a clean sweep was made of many just rights of property, as well as of much that was bad and obsolete. The ultimate fruits of the liberation of the soil were great and beneficial in the highest degree, but the

Sudden abolition of the feudal burdens, August 4, 1789.

immediate results may be easily guessed. The excesses of the peasantry were not lessened by the sudden annihilation of the bonds of ages; and they were only put down or checked at last by the efforts of the middle classes in the country districts, alarmed at the evident progress of anarchy. These, too, thus found themselves with arms in their hands, and almost independent of any kind of rule.

October 5 and 6, 1789. Such was France in August and September, 1789, old authority falling on all sides, power being transferred into new hands, and want and disorder felt everywhere, although for the moment restrained. The Court party meanwhile, scotched, but not killed, had been rearing its head at Versailles; and rumour spread that a band of loyal nobles were about to take the King to Metz, and to liberate him from 'rebellious subjects.' Troops, too, were gradually moved from the frontier; and the new National Guard at Versailles—for such a body had been organised—was treated with scorn and contempt at the palace. A sentiment had been growing up in Paris, and found favour in the Assembly, that the King should be removed to the capital; and the feelings of the masses, irritated by want, had become ready for any sudden outbreak. A scene, which occurred in the first days of October, became the signal for a new explosion of passion. A party of young officers, at a banquet in the palace, dashed down the Tricolour from their helmets in the presence of the

King and of his Court, at the sound of a well-known
royalist air; and, heated with wine, and lured by the
glances of courtly beauty and syren grace, vowed
that they, at least, would not abandon the Throne.
This second 'orgie' gave rise to a remarkable demonstration from Paris, though it is not easy to say
who were its chief designers. On the morning of
the 5th a procession of women, stung with hunger,
burst into the great Town Hall, and thence streamed
over the short space which separates the city from
Versailles, followed by savage and menacing crowds,
and ultimately by Lafayette and his National Guard.
The procession forced its way into the National Assembly, then discussing an unfavourable message from the
throne, and a party of these strange visitors was allowed
to enter the courts of the palace and parley with the
King. Order was restored when Lafayette arrived,
and the assemblage dispersed, to find exit as it
could, most of the soldiers having given it a welcome,
and the Body-guard of the King alone, a select
detachment, having provoked ill-will. Early next
morning a few chance shots, which struck down, unhappily, one or two of the people, became a for-
runner of a general rising; and a furious mob fell
on the Body-guards, and penetrated the interior of
the palace. Dread faces of passion, hunger, and
crime, appeared in the sanctuary of the State: the
Queen, half-clad, was driven from her chamber
amid the shrieks of affrighted attendants; and a

terrible massacre would have taken place but for the interposition of the late French Guards, who shouting 'We do not forget Fontenoy,' rescued the Body-guards and the royal family. A seeming reconciliation took place afterwards; the King presented himself from the balconies; the Queen gave her hand to Lafayette to kiss, and the Tricolour shone on every armed crest; but the floors of the palace were drenched with blood, and two ghastly heads, borne aloft on pikes, attested the presence of still unslaked passions. At the request of a deputation, peremptory though bland, Louis consented readily to go to Paris; and the royal carriages, with the King and Queen, their children, and Madame Elizabeth, the fair and pious sister of the King, slowly trailed to the city escorted by a roaring chaos of armed bands, of women astride on patriotic cannon, of savagery in its hideous or grotesque aspects. The

The King and Royal Family taken to Paris from Versailles.

shout, 'We have now the baker to ourselves, the baker's wife, and the baker's boy,' significantly told what thoughts were uppermost in the hearts of the poorer mass of the multitude—by some conspiracy they had been deprived of bread by 'aristocrats at Versailles.' It was evening before the motley procession made its entry into the gates of the Tuileries; and when the royal party reached the palace, uninhabited by the House of Bourbon for years, they saw themselves surrounded by National Guards, and were told that regular soldiers could not approach.

The events of these momentous days, known emphatically as the 5th and 6th of October, have been attributed to different persons; but it is superfluous to enquire whether Mirabeau,[1] the Duke of Orleans, or Lafayette, had any part in preparing the movement. What is to be noted is, that the rabble of Paris, though still controlled by the middle classes, had gained a great and marvellous victory; royalty had, as usual, shown itself ignoble, vacillating, and amiably weak; and the illusions of power, once feared and august, had been dissipated like the idlest of dreams. Since that day Versailles has been a national museum, and for a time a ruin; it has sheltered legions of German invaders, and heard the wailing cry of a conquered Nation; but never again has it been the abode of a Prince wielding the sovereignty of France.

Growing power of the rabble of Paris.

[1] It is now tolerably well ascertained that the Duke of Orleans instigated the mob to leave Paris, and attack the Palace. In a letter discovered after his death he directed a banker not to pay the money which had been agreed on as the price of the blood of the King. ' L'argent,' so he wrote, ' n'est point gagné, le marmot vit encore.' Mirabeau and Lafayette seem to have been innocent.

CHAPTER III.

THE CONSTITUTION OF 1790-1.

Character of the period from the autumn of 1789 to the summer of 1791.

THE next phase of the French Revolution may be fitly compared to the watery space, comparatively level, yet broken and tossed, and agitated by uncertain currents, which is seen occasionally between mountainous waves during the pauses of a tremendous storm. From the autumn of 1789 to the summer of 1791—a period of nearly two years—no events occurred of such obvious significance as the rising of Paris, the siege of the Bastille, the insurrection in the Provinces, and the 5th and 6th of October; and though elements of trouble gathered and grew apparent to a discerning eye, they did not yet form into a general outbreak. Some terrible crimes were, indeed, perpetrated under the influence of local passion or revenge; one or two conspiracies, real or feigned, were attempted by partisans of the Court; the emigration of the Nobles increased; all along the frontier rumours were heard of counter-revolution and even of invasion; the attitude of foreign Powers became doubtful; and throughout France, from the Pyrenees to the Rhine, innovation

showed itself in a thousand forms ; the hearts of men
throbbed with the desire for change, and knavery and
ambition appealed too successfully to the hearts, the
jealousies, and the fears of the multitude. Nevertheless, the surface of things at least appeared for a
time less disturbed than before ; some reforms were
attended with permanent good, and others with
benefits for the moment ; the dangerous pressure of
popular distress, so evident in 1789, lessened ; and it
seemed to thousands as if the Revolution was tending to happiness, peace, and progress. The King
had been separated from the faction of the Court ;
the National Assembly was supreme ; the removal of
the feudal burdens from the soil improved agriculture as if by magic ; France enjoyed such liberty as
she had never enjoyed before ; the Middle class and
the National Guards seemed sufficient to keep mob
violence down ; signs of increasing opulence were
not wanting ; and though disorder was still abroad,
and demagogues held formidable sway, and the
echoes of strife and discord were heard, were not
symptoms like these inevitable at a crisis of great
and rapid change, and would they not before long
disappear ? The issue was to be otherwise ; and this
brief moment of comparative calm was to see France
brought nearer to the abyss, to accelerate the dangers
collecting around, and ultimately to give renewed
force to revolutionary passion and suspicion. Yet
History rejects the false creed of fatalism, though

she admits the stupendous power of circumstance;
and while large allowances must, in justice, be made
for inexperience and the difficulties of the situation,
it is not the less, in our judgment, true that had
France found statesmen among her rulers, had her
aristocracy been less spoiled by arrogance, and less
morally worthless, had her Sovereign and those
around him been less unwise, the course of events
would have been very different.

The National Assembly begins to frame the Constitution.

After the scenes of the 5th and 6th of October,
the National Assembly returned to the task of re-
modelling the institutions of France, which had
been, almost from the first, its mission.   Much that
it accomplished during the following months,
although done with precipitate haste, was a great
improvement on the old state of things, and has
since had beneficent results.[1]   Old barbarous penal-
ties were abolished; seignorial jurisdictions disappeared; internal trade, which had been crippled
by mischievous restrictions, was set free; a project
was formed to fuse into a Code the medley of written
laws and customs, conflicting and obscure, which prevailed in the kingdom; the monopolies and exclusive
guilds of the towns vanished with the feudal charges,

Useful reforms.

---

[1] A learned account of the Constitution of 1790-1 will be found
in Professor Von Sybel's *History of the French Revolution.* Burke's
Reflections on the Revolution in France, remain, however, the best and
most profound commentary on the work of the National Assembly
and its tendencies. Many of the observations of the great philosophic
statesman have proved prophetic.

on the land ; and, above all, religious toleration was proclaimed, the whole system of taxation was re- formed, and the iniquitous exemptions of the pri- vileged orders in this particular were removed. These measures, and many others of the kind, were salutary, and, for the most part, just ; and English- men may, in these respects, agree with those French- men who extol ' the immortal principle of 1789.' But the work of the Assembly, considered as a whole, was marked by a passion for mere theory, and a perilous disregard of facts ; and it displayed itself in wild innovations which irritated and exasperated many classes, made settled government at best pre- carious, and added strength to revolutionary tenden- cies.   Instead of addressing itself simply to mitigat- ing the political and social grievances of which such a multitude existed in France, it had begun its labours by a grand Declaration—at once imposing, dangerous, and untrue—of what it regarded as the Rights of Man ; and it proceeded to carry out these principles, more or less faithfully, in its subsequent legislation.   Like the feudal exactions, the immense property of the Church, and of a number of other corporations, was confiscated with a stroke of the pen ; and though compensation was promised to existing interests, it was to a great extent illusory. Soon afterwards titles of honour were suppressed, however dignified or historical ; places, offices, and privileges were abolished wlth little reflection, or

Wild and precipitate innovations.

The Rights of Man.

Confisca- tion of Church and corporate property. Abolition of tithes.

even justice; and it was announced as an eternal
Equality. truth that Frenchmen were essentially equal, notwithstanding the inequalities that must exist in an
ancient, or indeed in any community. In addition,
an extraordinary change was made in the local con-
Provinces stitution of the Kingdom; the old Provinces were
trans-
formed into effaced from the map, with their complex variety of
Depart-
ments. rights and immunities; the local associations thus
formed were destroyed, and the Kingdom was par-
celled out into new divisions, ever since known by
the name of Departments, each with a perfectly
uniform organisation and distribution of local
authority.

Character Having thus levelled, in a few months, almost
of the new
Constitu- every institution of old France, the Assembly began
tion; its
glaring the work of creating a new Constitution for the
defects. transformed Kingdom. The Monarchy was continued and liberally endowed; but it was shorn of
most of its ancient prerogatives, and reduced to a
very feeble Executive; and while it obtained a
perilous veto on the resolutions and acts of the Legislature, it was separated from that power, and placed
in opposition to it, by the exclusion of the Ministers
of the Crown from seats and votes in the National
Assembly. The Legislature was composed of a
Legislative Assembly, formed of a single Chamber
alone, in theory supreme, and almost absolute; but,
as we have seen, it was liable to come in conflict
with the Crown, and it had less authority than might

be supposed, for it was elected by a vote not truly
popular, and subordinate powers were allowed to
possess a very large part of the rights of Sovereignty
which it ought to have divided with the King. This
last portion of the scheme was very striking, and
was the one, too, that most caused alarm among
distant political observers. Too great centralisation
having been one of the chief complaints against
the ancient Monarchy, this evil was met by a radical
reform, which also fell in with the new doctrines of
equality and the supremacy of the people—two main
tenets of the Rights of Man. The towns received
extraordinary powers; their municipalities had complete control over the National Guards to be elected
in them, and possessed many other functions of
Government; and Paris, by these means, became
almost a separate Commonwealth, independent of
the State, and directing a vast military force. The
same system was applied to the country; every Department was formed into petty divisions, each with
its National Guards, and a considerable share of
what is usually the power of the government; and
in each Department a higher administration was entrusted with a kind of general superintendence. In
every separate centre in town and country this immense authority was for the most part wielded by
men chosen by a scarcely restricted suffrage; and
Burke's saying was strictly correct, ' that France was
split into thousands of Republics, with Paris pre-

E

dominating and queen of all.' With respect to other
institutions of the State, the appointment of nearly
all civil functionaries, judicial and otherwise, was
taken from the Crown, and abandoned to a like
popular election; and the same principle was also
applied to the great and venerable institution of the
Church, already deprived of its vast estates, though
the election of bishops and priests by their flocks in-
terfered directly with Roman Catholic discipline, and
probably, too, with religious dogma. As for the
Army, it was also in a great degree removed out of
the hands of the King; and while unjust privileges
were swept away, it was organised on a democratic
model, commissions and similar rights being
abolished.

Administrative measures.

Many acts, too, of the National Assembly in the
administration of affairs were unwise and dangerous.
Notwithstanding the opposition of Necker, who,
though hardly a statesman, understood finance, it
was resolved to sell the lands of the Church to pro-
cure funds for the necessities of the State; and the
deficit, which was increasing rapidly, was met by an
inconvertible currency of paper, secured on the lands
to be sold. This expedient, borrowed, to some extent, from precedents set by the old Monarchy, and
indeed by other governments in distress, and not
wholly mischievous under careful restrictions,
was carried out with injudicious recklessness.

Assignats.

The Assignats, as the new notes were called, seemed

a mine of inexhaustible wealth, and they were issued
in quantities which, from the first moment, disturbed
the relations of life and commerce, though they
created a show of brisk trade for a time. In matters   False
of taxation the Assembly, too, exceeded the bounds   system
of reason and justice; exemptions previously en-   of taxation.
joyed by the rich were now indirectly extended to
the poor; wealthy owners of land were too heavily
burdened, while the populace of the towns went scot
free; and, though little wrong was as yet done, the
example was set of future injustice. Very large
sums, also, belonging to the State, were advanced to   Undue
the Commune of Paris, now rising into formidable   favour
power, at an interest much below the market rate;   shown to
and thus the Nation was made to minister to the   Paris.
needs of one favoured portion of it—a perilous and
iniquitous principle. With the assent of the Assembly, the funds so obtained were lavisly squandered in
giving relief to the poor of the capital in the most
improvident ways—in buying bread dear and re-
selling it cheap, and in finding fanciful employment
for artisans out of work. The result, of course, was
to attract to Paris many thousands of the lowest
class of rabble, and to add them to the scum of the
city; and, indeed,[1] not a few of the communistic
theories which predominated during the Reign of

---

[1] Professor Von Sybel's *History*, book II., chapter iv., has brought
out clearly the Communistic tendencies of part of the legislation of
the National Assembly.

Terror, and have ever since been a curse to France, may be traced, partly in operation, at this time.

The Feast of the Federation, July 14, 1790. Enthusiasm in Europe.
It is easy to point out on what erroneous principles the Assembly founded a large part of its work, and time was soon to show what a series of ills inevitably resulted from much that it had done. But the attractive nature of the doctrines it laid down, and the generous liberality of many of its speakers, created enthusiasm for the moment; and the declaration of the Rights of Man aroused feelings of exultation and delight, not only in France, but throughout Europe. On the first anniversary of the fall of the Bastille, and before the Constitution had been finished, Paris witnessed a scene which vividly expressed the sentiments with which millions welcomed what seemed the inauguration of a new age of gold. A great national holiday was kept; and, amidst multitudes of applauding spectators, deputations from every Department in France, headed by the authorities of the thronging capital, defiled in procession to the broad space known as the Field of Mars, along the banks of the Seine. An immense amphitheatre had been constructed, and decorated with extraordinary pomp; and here, in the presence of a splendid Court, of the National Assembly, and of the municipalities of the realm, and in the sight of a great assemblage surging to and fro with throbbing excitement, the King took an oath that he would faithfully respect the order of things that was

being established, while incense streamed from high-
raised altars, and the ranks of seventy thousand
National Guards burst into loud cheers and triumphant music ; and even the Queen, sharing in the
passion of the hour, and radiant with beauty, lifted
up in her arms the young child who was to be the
future chief of a disenthralled and regenerate people
—unconscious happily of the dark clouds that were
gathering already over so many victims. The following week was gay with those brilliant displays
which Paris knows how to arrange so well; flowery
arches covered the site of the Bastille, fountains ran
wine, and the night blazed with fire ; and the far-
extending influence of France was attested by enthusiastic deputations of ' friends of liberty ' from
many parts of Europe, hailing the dawn of an era of
freedom and peace.

The work, however, of the National Assembly
developed some of its effects ere long. The abolition
of titles of honour filled up the measure of the anger
of the Nobles ; the confiscation of the property of
the Church ; above all, the law as to the election of
priests, known as the Civil Constitution of the
Clergy, shocked all religious or superstitious minds.
The reduction of the old rights of the Crown, and
the antagonism created by the absurd severance of
the Legislative and Executive powers, enfeebled the
State, and caused the King, his Ministers, and the
Assembly to clash ; and Necker and all his associates

*[sidenote: Evil con-
sequences
of innova-
tion ; signs
of disorder
and
anarchy.]*

but one were dismissed, and replaced by men of an
inferior stamp. The extinction of privilege, too, in
the Army, provoked discontent among the whole
class of officers; and yet it did not much please the
men, for no great immediate benefits followed, and
their superiors stood more than ever aloof. Meanwhile the substance of power began to pass to the
masses to an alarming extent, through the regula-
tions as to the National Guards and the administrative services of the Kingdom; and though they did
not yet know their strength, leaders were not wanting to teach the lesson. The ascendency of Paris,
too, became more decided than it had ever been;
and the dislocation of authority caused by the extreme weakening of the central government dis-
integrated France in a great degree, and gave a wide
scope to low popular influence. It is easy to imagine
the results, in a country torn by deep divisions of
class, where an ancient throne had been suddenly
weakened, where nothing was permanent, fixed, and
established, and where anarchy and license, though
for the moment checked, had made themselves so
perilously apparent. The emigration of the Nobles,
which had become very general from the 5th and 6th
of October, went on in daily augmenting numbers;
and, in a short time, the frontiers were edged with
bands of exiles breathing vengeance and hatred. In
many districts the priests denounced as sacrilege
what had been done to the Church, divided the

peasantry, and preached a crusade against what they called the atheist towns; and angry mutinies broke out in the Army, which left behind savage and relentless feelings. The relations between the King and the Assembly, too, became strained, if not hostile, at every turn of affairs, to the detriment of anything like good government; and while Louis sunk into a mere puppet, the Assembly, controlled in a great measure by demagogues and the pampered mobs of Paris, felt authority gradually slipping from it. Thus anarchy was not restrained from above, while, so to speak, it was organised from below; and the rein was thrown on the necks of a people long misgoverned, and whose excitable nature had been aroused by every kind of stimulant. As yet, however, the mere popular forces did not break out in general disorder; but their increasing influence was plainly seen in the ascendency gained by brawling demagogues, in an immense diffusion of cheap and bad journals, and in the multiplying of associations of an extreme type in politics. One of these societies, sprung from a small beginning, had established itself in an old convent in Paris; and here, growing into larger numbers, it held frequent sittings, at which the members discussed the acts of the National Assembly, or made vehement addresses to the people. The most ardent reformers of the Commune were prominent in it, and were wont to report to the populace, in the forty-eight sections into which the

The Jacobin and other clubs.

capital had been divided, whatever had been decided or done ; and the society had affiliated to it a great number of bodies of the same character throughout the principal cities of France.   Such was the origin of the famous Jacobin Club—a dread name in the drama of the Revolution.

Weakness of Conservative elements in the Assembly.

It may appear strange that the powerful interests which were represented in the National Assembly did not contend better against these immense changes ; and that the Commons, of whom very few had genuine sympathy with the lowest classes, should have given such free scope to anarchic disorder.   But the Crown and the Nobles were divided from each other ; the Nobles were divided among themselves ; the prelates and lower clergy were not friends ; and many of the lay and clerical aristocracy were unwise enough to join the ranks of the emigrants.   Of the Conservative Nobles and prelates who remained in the Assembly, few had anything like talent ; and the chief defenders of the ancient rights of the throne of Henry IV., and of the Rohans and Mortemarts, were a young dragoon officer and a simple abbé, the impetuous Cazalés,[1] the subtle Maury.[2]   As for the

[1] Cazalés, the brilliant military champion of Conservatism in the Assembly, was born in 1752.   He has been well described as 'a chivalrous soldier, sans peur et reproche ;' and Mirabeau said of him that 'if the knowledge of Cazalés equalled the charms of his elocution, all efforts would be ineffectual against him.'   But he was rash to a fault, and seems to have had as little judgment as information.

[2] The Abbé Maury was born in 1746.   He had been versed in

reforming Nobles, among whom were several men of
fine parts, many doubtless went further than they
wished ; but some were carried away by the false
philosophy in fashion ; others bid against each other
for popular support, and they never united in a
rational policy of what might have been called constitutional reform. The Commons, too, were mere
tyros in politics, though many were apt at the
tongue and pen ; they were also full of the new doctrines, and could not see what innovations were un-
safe ; and they were largely influenced by a strong
dislike of the old institutions and the privileged
orders. Add the characteristics of the French intellect, addicted to system, and to carry out ideas,
without regard to facts, to their extreme consequences ; add the impetuous and ardent French
temperament, often wildly generous and sentimental ;
and we shall see how the Assembly, without any intention, prepared the way for a part, at least, of
what followed. Yet what contributed most, per- .
haps, to the annihilation of the noble classes, and
encouraged measures of a revolutionary tendency,
was the pitiful conduct of those best known by the
still dishonourable name of *émigrés*. In a few
months the great majority of the aristocracy of

ecclesiastical and political affairs before the Revolution, and defended with skill and eloquence the cause of the Monarchy, the
Church, and the Nobles in the National Assembly. He became
afterwards an Archbishop and a Cardinal, and died in 1817, having
witnessed the Bourbon Restoration.

France had fled the kingdom, abandoning the throne around which they had stood, breathing maledictions against a contemptuous Nation, as arrogant as ever in the impotence of want, and thinking only of a counter-revolution that would cover the natal soil with blood. History makes allowances for these men, for they were the victims of an evil order of things; but France could not make allowances for them at a crisis of agitation and passion; and their utter want of patriotism and of sound feeling made thousands believe that the state of society which had bred such creatures ought to be swept away.

Attempts of Mirabeau to check the disorganisation of the State.

One man, however, in the National Assembly, saw distinctly whither events were tending. The life of Mirabeau was stained with vices; and his public career was deeply marked by reckless ambition and perhaps crime. But he added keen insight and strong common sense to eloquence of extraordinary force; he was not the dupe of deceptive theories, and he perceived that France was falling into confusion. He had protested against the destruction of the Church and Nobles as leading to civil war; he had declared that it was dangerous and unwise to refuse the ministers of the Crown a seat in the Assembly; and he summed up a great truth in the words that what France required was a firm Executive to keep anarchy down and to maintain order. We cannot affirm whether he had thought out a scheme of Constitutional Monarchy for France; but

as early as 1790 he made overtures to the Court, and
he had more than one interview with the Queen, to
whose ' force of character' he did admiring homage.
His projects were to remove the King to a town in
the interior of France, to rally around him the loyal
part of the Army, and to summon a new Assembly,
which would undo what was most mischievous in the
work of the old ; and he promised that he would
answer for thirty-six Departments, and expressed a
strong hope that the middle classes, alarmed at the
prospect of mob rule, would throw their weight on
the side of the Monarchy.   The Court, however,
vacillating and suspicious, would not trust the proud
man of genius ; and Mirabeau could not obtain the
adhesion of Bouillé, the most popular chief of the
Army, and of Lafayette, all powerful with the
National Guards, whose co-operation he deemed
necessary.   Death came to put an end to his hopes   His death.
and fears ; he expired in April, 1791, and with him
perished the best chance of arresting the Revolution
already at hand.

Meanwhile, the attitude of neighbouring States   Threatening attitude
had become uncertain, if not threatening, and   of Foreign
Powers.
sounds of counter-revolution, and even of war, had
begun to gather definite shape.   The old Monarchies
and Aristocracies of Europe were naturally alarmed
at what they called French principles ; and Prussia
and Austria suddenly composed their feuds, while, in
England, the House of Commons rang with cheers at

Burke's invectives against the Assembly. The *émigrés*, too, were in every Court soliciting aid and making empty noise; and a little Army of Seigneurs formed along the Rhine, which they boasted was only the advanced guard of a crusade in their holy cause. The death of Mirabeau at this conjuncture made the King and Queen despair of obtaining deliverance through French help alone; and they began wistfully to look abroad, and to dabble in those foreign intrigues which were to end in the destruction of both.[1] Both, indeed, feared and disliked the *émigrés*; nor did either, as yet at least, think of restoring the Monarchy by foreign bayonets. But, not to speak of many other grievances, the conscience of Louis was wounded to the quick by what he thought was sacrilege done to the Church; Marie Antoinette resented the eclipse of the throne, and slights offered to her consort and herself; and the condition of France appeared to both alike something unintelligible and beyond endurance. A project was formed that the King should escape,

Dangerous projects of the King and Queen.

[1] A complete picture of the dealings of Louis XVI. and Marie Antoinette with Foreign Powers during the Revolution, and indeed of the life and conduct of the Royal Family at this crisis, will be found in the collection of *Letters of Louis XVI., Marie Antoinette, et Madame Elizabeth*, edited by M. Feuillet de Conches. Some of these documents may be apocryphal, but enough of undoubted genuineness remain to show that the suspicions entertained by the popular leaders of the King and Queen were in part justified. The collection throws also much light on the policy of the Continental Courts from 1790 to 1793, and the dissertations and notes of the editor are valuable and interesting.

should place himself in the hands of an Austrian
detachment, to be marched quietly within the frontier, and should make an appeal to all loyal Frenchmen ; and no doubt can remain but that a violent
change in the Constitution was in contemplation to
be effected by the ' friends of the Monarchy.' The
idea of calling in Austrian troops shows that the
Queen, a daughter of the House of Austria, was the
chief author of this perilous scheme ; and Bouillé,
who was in command in Lorraine, was let into the
secret, and promised assistance. On the night of The flight
June 20, 1791, the Royal Family set off, eluding to Varen-
the guard at the gates of the Tuileries, the King 20, 1791.
having left a declaration behind, in which he disavowed all that had been done in his name, and
with his assent, since ' he had been in durance in
Paris,' and denounced the Constitution he had sworn
to maintain. The carriage arrived safely next day
at Châlons, but was stopped at last at the little
town of Varennes, a postmaster called Drouet having
recognised the King, and given the municipality a
hasty warning. A parley took place, the ill-fated
Monarch, as usual, showing ignoble weakness ; but
the party would have escaped had not a detachment
sent by Bouillé shouted ' the Nation for ever,' and
refused to obey their officers' orders. Meanwhile, at
the news of the flight of the King, the Assembly
decreed the recall of the fugitives, and assumed the
functions of complete sovereignty, and commis-

sioners were despatched to enforce its mandates. On the arrival of the delegates at Varennes, Louis instantly yielded, in spite of the entreaties of the Queen ; he seems, it is said, ' to have been most anxious about finishing his morning meal.' The Royal captives were eight days returning, every village looking on at the sorry sight; and the procession threaded the streets of Paris amidst a multitude silent, and with covered heads. Pétion, one of the commissioners, had been rude and forward ; another, Barnave, had been fascinated by the Queen, and had shown that his feelings had been deeply touched. Each of these men were to go different ways in the dark drama of future events, in consequence, perhaps, of this accident in their lives.

Consequences of this event.

Napoleon said, many years afterwards, that this abortive attempt sealed the fate of the Monarchy ; it at least caused general indignation and distrust. The Nation did not, indeed, know the whole truth ; but the protest of the King against the constitution was read by many in the worst light, and what was really feebleness was thought treachery. The demagogues swarming in France made the most of the chance which had fallen to them ; angry meetings were ·held in many places, and appealed to in threatening and wild language ; the vile name of Marat [1] emerged from darkness, and those of Dan-

---

[1] Jean Paul Marat, born in 1744, was bred a physician, and afterwards became a veterinary surgeon in the stables of the Count

ton,[1] Robespiere,[2] and others, soon to become by-
words of universal terror, were repeatedly in the
mouths of the populace, and even the word 're-

of Artois. Though unsuccessful in his profession, he was not without parts, and at the beginning of the Revolution he became editor
of the *Ami du Peuple*, one of the most violent journals of the time.
In this evil production he systematically advocated the destruction
of the upper and middle classes, and the subversion of property.
Though at first decried even by the demagogues, he by degrees
emerged from obscurity, and became one of the most prominent and
repulsive figures of the Reign of Terror. He was assassinated in
1793.

[1] George Jacques Danton was born in 1759. This remarkable
man was brought up to be a lawyer, and plunged with characteristic
energy into the vortex of the Revolution. Mirabeau soon discovered
his talents, and he quickly became the most effective of the mob
leaders of Paris. He rose to be the most conspicuous actor in the
Revolution during the first part of the Reign of Terror; and though
his crimes were many, his courage and patriotism plead for him.
He laboured also, at the risk of his life, to reconcile parties, and to
stop the tyranny of the Committee of Public Safety, and was guillo-
tined in 1793.

[2] François Maximilian Robespierre was, like Danton, born in
1759, and bred an advocate. He was one of the few extreme Revolutionists who obtained a seat in the States-General; and for some
months he could not get a hearing when he attempted to speak,
though Mirabeau predicted that his earnestness would raise him into
notice. He became one of the chief of the Parisian demagogues,
winning his way, not by eloquence or boldness, but by a reputation
for integrity, and ultimately stood at the head of the most frightful
Dictatorship Europe ever saw. Opinions are still divided whether
he was a mere bloodthirsty tyrant or a merciless fanatic; I incline
to the second view. Mr. Carlyle has thus graphically described
Robespierre:—'Who of these Six Hundred may be the meanest?
Shall we say that anxious, slight, ineffectual-looking man, under
thirty, in spectacles, his eyes, were the glasses off, troubled, careful;
with upturned face, snuffing dimly the uncertain future times; com-
plexion of multiplex atrabiliar colour, the final shade of which may
be the pale sea green?'

public' was heard in the Assembly. That body,
besides, had pronounced itself the Government
during the flight to Varennes; the King and Queen
were so completely prisoners that they could hardly
leave the courts of the Tuileries; and captive royalty
seemed dead to the multitude. Nevertheless order
was still maintained, at least to a considerable extent;
some of the mob-leaders in Paris were silenced for a
time; and an attempt at an outbreak was put down
by Bailly and Lafayette, not without blood. Signs
of returning confidence, too, were now and then
seen; Louis and his consort were more than once
greeted with cheers when allowed to show themselves at the opera; and, notwithstanding all that
had taken place, generous hearts felt for their fallen
splendour. A kind of instinct, also, told the middle
classes that an hour of trial and peril was near; and
at the distant sound of the approaching tempest
thousands turned their eyes towards that tottering
throne, which, burdened as it was with evil memories,
seemed at least a rallying point against anarchy.

CHAPTER IV.

THE LEGISLATIVE ASSEMBLY.

Wᴇ have now reached a period when the elements of disorder, comparatively quiescent for a time in France, were to burst out with a force never seen before; when the consequences of her ancient ills and discords, and the vices of her new made institutions, were to reveal themselves with portentous distinctness; and when popular passion, given sudden strength in the weakness of an ill-organised State, and quickened by the most powerful incitements, was to overthrow the already unstable Monarchy, and to lead ere long to a terrific catastrophe. These events, fraught with momentous issues, were again to give proof how deep-seated were many of the mischiefs afflicting the kingdom, were again to bear witness to the want of wisdom of the higher orders and the unhappy Sovereign, were to bring out again in too clear relief the faults of the work of the National Assembly, and the numberless resulting perils and evils; and were, too, to show by a fresh awful example how power may suddenly fall away from classes which seemed in possession of it; and

Character of the period from the summer of 1791 to August 10, 1792.

F

how, in France especially, perhaps, a vehement, reckless, and daring minority, may, in certain circumstances, overbear opposition, and rise into powerful and commanding authority. The causes, however, of these phenomea were not to be sought within France alone; they were largely due to the influences external to her; and but for elements alien from it in the main, and arousing national excitement to frenzy, the course of the Revolution would, we think, have been different.

Deceptive calm; meeting of the new Legislative Assembly.

A few weeks after the flight to Varennes the National Assembly dissolved itself, having declared that its great work was done, and that it left ' France, as it hoped, regenerate.' The King, having with apparent readiness accepted the Constitution he had laboured to subvert, and a general amnesty having been proclaimed, the political sky seemed for the moment brightening; and the royal family were seen in Paris in splendour, and what it was assured was liberty. The new Legislative Assembly met in October; and, as the elections had been tolerably quiet, many began to hope that ' the glorious gains' of the two preceding years were at last realised, and that what had been done was finally established. One circumstance, however, was not in favour of the political experience of the fresh made Legislature. The National Assembly, by a self-denying ordinance, had excluded its members from its successors' ranks; and thus the Representa-

tive Body of 1791 was as ill-trained as the States-
General had been. The Legislative Assembly, of
course, contained very few members of the late
Noble Order, and was chiefly composed from the
Middle classes; but its prevailing wish was, at first
at least, to uphold the constitution of 1790–1,
though a republican party existed in it, and it was
desirous of trying its hand at reform. It num- Character
bered 750 deputies, divided into a Right, a Centre, Body. of this
and a Left party; the first conservative, the last
radical, and the second, which was very numerous,
wavering between them. The Left, in the political
slang of the day, became known as the Mountain,
and the Centre as the Plain; and these odd nick-
names were, before long, to acquire a strange
and world-wide celebrity. Taken altogether, the
Legislative Assembly was inferior in brilliancy to
its predecessor; and it had no member of the
powers of Mirabeau. It was not, however, without
men of fine parts; and one knot of deputies, known
afterwards by the name of the Department from
which many of them came—the Gironde, part of the The Gi-
old Province of Guyenne—became conspicuous for ronde.
their persuasive eloquence, and their ardent, if not
very wise, enthusiam.

For a brief space the Assembly and the King Dissensions
made a sincere effort to work well together. Louis between the
King and
composed his ministry of moderate men, known as the Assem-
Feuillants, from a club of that name, set up to bly.

F 2

counteract the power of the Jacobins; and some of
the reformers in the National Assembly, among
whom Barnave was conspicuous, endeavoured to
aid the Court with their counsels. From the first
moment, however, the King had been vexed by
slights put on him by some deputies in return for
want of courtesy on his part; irritation was aroused
by what was thought an extravagant augmentation
of the military force allowed the Sovereign by the
Constitution, and by the refusal of Court lords and
ladies to accept places in the newly-arranged household; and causes of dissension quickly multiplied,
and were aggravated by the absurd provision which,
by excluding them from seats in it, made the Ministers of the Crown appear dictators and strangers to
the Assembly. Meanwhile the acceptance of the
Constitution had stirred the *émigrés* to increased
wrath; and though their army as yet made no signs
of life—for noble officers could get few soldiers, and
would not stoop to serve in the ranks—the brothers
of the King uttered vehement protests against the
acts of the National Assembly, and their emissaries
swarmed across the Rhine, preaching discord and
Disorders in the Provinces. trying to excite trouble. The movement was largely
seconded by the priests, who had very generally
refused to take an oath to respect the new order of
things in the Church; and it created not a little disturbance, ancient sympathies and religious feelings coming into collision, in many Departments,

with the sentiments and instincts of the Revolution. Religious and social troubles. Massacre at Avignon (Oct. 16–20, 1791).
Some frightful disorders, too, broke out at Avignon,
for ages a fief of the Pope, but annexed by the
National Assembly to France. The French and
papal parties rose against each other, and a foul
and hideous massacre took place; several towns in the
South and South-east were convulsed by the strife
of angry factions already approaching a civil war;
and bands of armed and houseless vagrants prowled
about the kingdom in many districts, levying black
mail, and doing all kinds of outrage. A feeling of
general discontent, besides, grew visible ere long in
the great cities, and the populace was occasionally
on the verge of insurrection. The assignats had
quickened trade at first, but the reckless use of paper
money produced before long the inevitable results;
and the prices of everything rose rapidly, while,
in the unsettled [position of affairs, employers became alarmed and cautious. The millions of artizans and workmen felt the necessaries of life grow
dearer and dearer, while their wages did not increase
in proportion, and there were many to tell them
that this was caused by the selfishness of the rich
and of jobbing speculators. Besides, the Revolution,
it was said, was over; yet, after all, what had it
done for the poor, how had it realized the fair
visions held before the multitude by specious orators?
The Middle classes were, no doubt, better off; but
the Middle classes were as hard as the nobles; and

was the end to be a mere change of masters? These dangerous sentiments were fanned and excited by the Jacobins in Paris and kindred societies; and the fires of agitation spread far and wide, at first smouldering and irregular, but gradually gathering volume and force.

Louis directly opposes the Assembly.

These disorders were not calculated to better the relations of the Assembly and the King, or to give repose to the national mind; and the elements of mischief soon became more active. The Assembly, at different times in the winter, passed severe decrees against *émigrés* and priests; and though these may have been unnecessarily harsh, the occasion justified strict precautions. Louis, however, possibly with a

Nov. 1791.

kindly sentiment, but certainly with extreme imprudence, affixed his veto to these measures, and thus directly opposed the will of the Legislature, although expressed in a decisive manner, and that, too, on a most important occasion. The grievance was aggravated by the fact that the King was continually in communication with his brothers, at least, in the *émigré* camp; and that, parading his dislike of the new settlement of the Church, he had chosen his chaplains and confessors from priests who had refused to take the constitutional oath. We can readily conceive the results of this folly—which Englishmen would not have borne for a day, though trained for centuries in political life—in the case of a Legislature without experience, and exceedingly jealous of its new rights, and of a nation vehement, ignorant, and distracted. The Assembly, hitherto inclined to act with the King, became full of anger and suspicious fears; many moderate men fell off from the Court; and the Legislature being for the time baffled, the general indignation found vent in passionate outbreaks of popular wrath. The Commune of Paris besieged the Assembly with petitions signed by its united sections; the Jacobin Club, and one even more violent, the Cordeliers, grew into increased importance; and the Jacobins regularly usurped the functions of a great deliberative body, and despatched emissaries to stir up the people in every part of the kingdom. By these means, aided by the affiliated clubs, and by a Press growing in rabid license, existing elements of discontent and anger were made intense, and 'a general outburst of passion was organised against the King, the Court, and even the Constitution, which it was asserted had proved worse than useless. Meanwhile the masses were everywhere taught to seize and make use of their immense power; the machinery of government and of administration fell more and more into the hands of the populace, under the superintendence and control of demagogues; and the National Guards became more and more filled with what were called patriots, or were debauched by them. This was more especially the case in Paris, where, Lafayette having resigned his command, the National Guard had been remodelled,

Indignation of the Assembly.

Attitude of the Commune of Paris. The Clubs and demagogues.

and where Bailly had been replaced by Pétion, a false and artful popularity seeker, though the Court, with hardly intelligible silliness and spite, had thrown its influence into the scale on his side for the office of mayor against Lafayette. By this time the National Guard of Paris had been largely recruited from the lower ranks; its officers had been in some degree changed; and it was divided in mind and general sentiment. It was still, on the whole, on the side of order; but its discipline had been much relaxed; and in the contest between the Assembly and the King its sympathies were with the popular cause.

Menaces of Foreign Powers.

It was thus faring ill with regenerate France, by reason of crime, misrule, and bad institutions. Meanwhile an influence was drawing near from without which was to give sudden and appalling strength to the elements of disturbance in the State, and to stamp a character on the Revolution more terrible than any it had yet shown. Not long after the flight to Varennes, the Sovereigns of Austria and Prussia had met; and, at an interview at the little town of Pilnitz, had laid down a plan for a Coalition against France, with a solemn protestation that the cause of Louis XVI. was that of all the Monarchies of Europe. German writers,[1] whose mission it has been of late

Declaration of Pilnitz, August, 1791.

---

[1] Professor Von Sybel is the most eminent of these apologists, but his arguments are mere special pleading; and the very fact that the Sovereigns of Austria and Prussia, hitherto hostile, issued the manifesto of Pilnitz, was more than enough to provoke France.

to deny the possibility of German ambition, may say that this League was a mere pretence; but it could not have so appeared to Frenchmen, and it was followed by demonstrations of force, and by fair words at least to the collection of *émigrés* who were menacing France from German territory. Prussia, too, at this time was hankering after Alsace; the Count of Artois had been base enough to hint at concessions in Lorraine as the price of aid to the good cause; and Russia, Sweden, Piedmont, and Spain, held an attitude more or less threatening. It is certain, moreover, that for months afterwards the King and Queen were in regular correspondence with the Emperor of Austria and other Sovereigns; and though they still repudiated foreign war, Marie Antoinette urged the necessity of an armed Congress upon the frontier, which would place the Monarchy on a new footing, and of course lead to a counter-revolution. These intrigues, however, were not fully known, though a kind of instinct apprehended part of the truth; and the Legislative Assembly was for a considerable time divided on the question of peace or war, the Moderates and the Gironde being for hostilities, the extreme Left condemning an appeal to arms as tending to a dictatorship and the rule of the sword. After an envenomed controversy with his opponents, the Austrian Minister, Cobentzel, at last indicated an intention on the part of his master to intervene directly in the affairs of France, and to put

Projects of the King and Queen.

Divisions in the Assembly.

down the Revolution by force, and this precipitated
the impending crisis.   The Assembly declared war
against Austria in April, 1792, the King assenting
with seeming readiness.   Prussia eagerly coalesced
with her Austrian rival; and thus commenced a
tremendous conflict, which was to shake the world
for twenty-three years, and in which France, we
think, was not the first aggressor.   Three French
armies were despatched to the frontier; but the
soldiers, spoiled by long neglect of discipline, were
unable to look in the faces of their foes, and they
were driven out of Belgium after a severe skirmish,
slaughtering in their fury one of their chief officers.
Before long, too, it became evident that the whole
forces of the Kingdom were demoralised, and wanting
in almost every appliance of war; and Lafayette,
who commanded one of the three armies, and in
whom confidence was generally placed, made no secret
of the danger of the situation.

These tidings spread consternation through
France, and exasperated the passions that stirred the
capital.   The war party in the Assembly, swayed by
the Gironde, at once acquired a decisive ascendency;
the leaders of the populace shouted treason, and
vociferated that the troops had been sold; and dark
suspicions accumulated against the Court, the King,
and especially the Queen.   Even in the Assembly it
was openly said that an Austrian committee sat in
the palace, betraying the dearest interest of France;

Margin notes:

It declares war against Austria. Prussia joins Austria.

French failures in Belgium. April, May, 1792.

Indignation in France.

and fierce threats were uttered that an Austrian woman should not be allowed to stand in the way of the Nation. Ere long the insensate conduct of Louis added fresh fuel to the kindling excitement. He had been compelled, before hostilities began, to part with his Feuillant administration, one of the members of which had been impeached for pusillanimity in the negotiations with Austria; and he had formed a ministry mainly composed from the Gironde and popular party, but of which the real chief was Dumouriez,[1] an able and brilliant soldier of fortune. This cabinet had proposed a new law against the non-juring and half-rebellious priests, and the Assembly had voted it with acclamation; and soon after the defeats on the frontier one of the ministers brought forward a measure for creating a camp of twenty thousand volunteers near Paris, which would become the nucleus of a national army to be drawn together from all parts of the kingdom. An enthusiastic vote welcomed this scheme also ; and it deserves notice that although Dumouriez expressed at first his dissent from the project, and made use of the opportunity to intrigue against three of his Gironde colleagues, he recurred to it almost immediately, however worthless and dangerous, besides, such a force must have seemed to an experienced

Insensate conduct of Louis.

[1] Dumouriez, born 1745. This able general and brilliant diplomatist had served and intrigued with distinction before the Revolution. His character is drawn with skill and fidelity by M. Thiers, *Histoire de la Revolution Française*, vol. ii. p. 58, ed. 1842.

He resists the decrees of the Assembly and dismisses the Gironde ministry and Dumouriez.

soldier. The King, however, even at this crisis, directly thwarted the vote of the Legislature ; refused to sanction the double decree ; dismissed first the three Gironde ministers, and Dumouriez himself a few days afterwards ; and chose a new Cabinet from the unpopular Feuillants, suspected, in part at least, of weakness, and discredited in the eyes of the Assembly and the Nation.

Outbreak of June 20, 1792.

History justly condemns the excesses that followed, and the bad use that was made of popular passion ; but neither ought she to forget the provocation, or the circumstances that led to the triumph of anarchy. The leaders of the Assembly, once more brought into collision with a Sovereign and a Court believed to be leagued with the national enemy, at a crisis of sudden national peril, turned to the capital for support ; and while they denounced openly the conduct of the King, they sought the aid of the demagogues and mobs of Paris as instruments against the intrigues of the palace. This course was unwise, and in part selfish, but motives of patriotism concurred ; nor is it perhaps surprising that these men made this wild appeal to revolutionary forces. Pétion, the Mayor of Paris, gladly organised the powers of the Commune to stir up agitation ; the Jacobins and Cordeliers called on all ·patriots to take up arms to resist oppression ; and the galleries of the Assembly were nightly thrown open to swarms of ferocious and squalid spectators, who clamoured down attempts

to oppose the majority. In a very short time the
streets of the city were once more dense with masses
of pikemen, who overawed or won over the National
Guards, and this growing army of savagery was
largely recruited from the desperadoes who for some
months had been congregating into the dens of the
capital. An occasion for an outbreak soon arose, and
there can be no doubt that it was at least connived
at by many in the Assembly and by the municipal
authorities. On June 20, the anniversary of the oath
of the Tennis Court, a great crowd collected to com-
memorate that event, and it burst armed into the
hall of the Legislature, waving banners with murderous or grotesque emblems, and calling on the
deputies to act with energy. The mass, unchecked
and even welcomed, next broke through the gates of
the Tuileries, and the courts of the palace were soon
filled with an excited multitude, crying, ' Down with
the veto,' ' The Nation and the patriot ministers for
ever.' Several thousand National Guards were present, but they looked on with indifference or had no
orders; and one battalion, it is said, shouted 'that
it knew who was its real enemy.' The chambers of
the royal family were quickly reached, and at the
sight of Madame Elizabeth yells arose fiercely
against the ' Austrian woman,' the princess being
taken for the Queen, while the King was assailed with
epithets of ' Monsieur Veto,' and ' the Constitution
or death.' The impassive attitude of Louis, how-

Encour-
aged by the
popular
leaders in
the as-
sembly and
by the
Commune
of Paris.

ever, had some effect in calming the crowd, and no
hand was lifted up against him, though a cap of
liberty was thrust upon his head, and he remained in
this humiliating position for hours, surrounded by
execration and ribaldry. The Queen, meanwhile,
had been happily rescued by the efforts of a few
courageous men ; and awestruck, it is said, by her
majesty and grace, her intending murderers turned
aside their weapons, while a few kindly words from
her lips melted into tears some of the female furies
who had hung on the skirts of the hideous procession.
Towards evening Pétion, who had at least offered no
opposition to the demonstration, persuaded the multitude to disperse ; but the secret of the defenceless
state of the palace had been discovered, and was not
forgotten ; and royalty seemed, as it were, trailed in
the dust.

Reaction in
favour of
Louis.

The disgraceful scene of June 20 caused a slight
reaction in favour of the King. The patience of
Louis excited compassion ; the Assembly began to
dread the forces which its leaders had rashly called
to their aid ; and the Gironde party, appalled at
the prospect, made overtures for the recall of the
three Gironde ministers. Lafayette, too, hastened
from the frontier to condemn the violence of the
Commune and the Jacobins. Pétion was prosecuted
for his conduct on the 20th, and the moderate citizens, still the majority, were sincerely desirous of
seeing order restored. But the movement ere long

made renewed progress, precipitated by the intelligence of fresh defeats, by passion, and by the obstinacy of the Court. On June 30 the Assembly passed a resolution that all existing authorities should be in permanent session, and thus the organisation of democratic forces which had been created all over France, and had fallen under the control of demagogues, was kept in motion to excite the people. Petitions began to pour in from the provinces ; the towns fermented with angry agitation ; the municipal assemblies, and those of the Departments, were mastered by low and reckless mobs, all more or less with arms in their hands, and Paris formed the centre from which this machinery was worked by those who managed those turbulent masses. Meanwhile an attempt was made to create the very armed force which the King had opposed ; volunteers were invited to flock to Paris for the approaching commemoration of the fall of the Bastille ; the Constitutional Guard of Louis was disbanded ; and the staff of the National Guard was changed and filled with men of a revolutionary type. At the same time the ferocious bands who had shown their power on June 20 were held in the leash by their desperate leaders, and vile incitements were not wanting to urge all ' patriotic men ' to join them. The Commune of Paris, almost independent and Sovereign within its own limits, was, in the main, responsible for those measures ; but the majority of the Assem-

New efforts of the Demagogues and the Commune.

bly concurred, and they were attended with the desired results.   On the day of the festival Louis found himself in the presence of a host of armed men—— many come from distant parts of the kingdom—who either maintained an ominous silence, or shouted, 'The Nation,' 'Pétion,' or 'Death;' and even the National Guards were wild and unsteady.   By this time the state of the capital had become so alarming that the King was implored by ministers and trusty friends to fly; two high-souled noblemen, faithful among the faithless, placed their wealth at his feet; and even Lafayette promised to come to his aid and to take him in safety to the army.   But irresolution and evil councils prevailed; the unhappy monarch refused to move; and Marie Antoinette exclaimed, in a burst of passion, 'that she would rather perish than trust such a hypocrite as Lafayette.'   Nor were other motives, as we now know, wanting: the King and Queen had been kept informed of the intended march of the German armies; and she had boasted exultingly that her deliverance was at hand.   Pity as we may an august victim, that deliverance would have been wrought in blood and fire, even if this result had been against her will; and truth requires us to note the circumstance.

Proclama-
tion of
Brunswick.
France and Paris were in this critical state when a memorable incident suddenly removed the last checks on the revolutionary forces.   At the end of July the Prussian army, under the Duke of Bruns-

wick, was set in motion, two Austrian divisions being on either wing; and the invading host, headed by bands of *émigrés*, wild with delight and thirsting for revenge, advanced from the Rhine to the Moselle ·and the Meuse. Brunswick issued a proclamation, ever to be condemned by those to whom national freedom is dear, and which years afterwards met its fitting reward. This manifesto, among other outrages, summoned Paris instantly to ' submit to its King,' declared that it would be ' razed to the earth ' if any insult were offered to the royal family ; and, after announcing that the ' Legislative Assembly, the National Guards, and the municipal authorities would be held answerable for whatever occurred, to military courts-martial, without a hope of pardon,' kindly added that ' their Austrian and Prussian Majesties would do their good offices with his most Christian Majesty to obtain forgiveness for his rebellious subjects.' This infamous document caused a thrill of fury and wrath to shoot through the capital; and though Louis, no doubt sincerely, disavowed what the Allies had done, the mischief, unhappily, was beyond recall. In the outburst of indignation which stirred the citizens, the first thought was of safety and vengeance ; and as the Assembly, at this crisis, did little but applaud the orators of the Gironde, and had no resolute and practical policy, power passed quickly to the more reckless dema-

G

gogues, and there was hardly anything to oppose the

most desperate projects, though the party of order was still the most numerous. An insurrection was regularly planned, its object being to dethrone the King, and to keep him a hostage with the rest of his family; and, as we have seen, means to work on the populace, and formidable armed power, were not wanting, while all other authorities were weak and doubtful. Revolutionary committees, as they were styled, were formed in the Jacobin Club and in the Commune; and delegates from these harangued the

sections, called upon them to organise themselves and rise, and laid the train for a general explosion. Danton shone eminent among these leaders; and his terrible aspect, fierce earnestness, and rude, savage, but genuine eloquence, had already gained him the name of the 'Patriot Mirabeau.' By this time thousands of volunteers had arrived to swell the bands of Parisian pikemen; and among them the contingent from Marseilles, 'six hundred men who could do or die,' were conspicuous for their audacious bearing. The rising was fixed for August 9; and as some of the members of the Commune were not willing to go the necessary lengths, it was resolved to replace this body suddenly by men of the true patriot stamp from the sections. Pétion, treacherous and timid, assented to the scheme, so that his hand in it should not be seen; and it was veiled under a show of legality, an immense petition from the forty-eight

sections for the immediate setting aside of the king
having been presented to the Assembly.

On August 9, when darkness had fallen, the
note of preparation began to sound. The summer
moon was calm in the heavens ; and all those who in
a great city love quiet, whatever the passions of the
hour, were sunk in sleep, unconscious of what was to
come. Many, too, though by nature friends of order;
also half knew that wild schemes were abroad, and
were not sorry that a stern lesson was to be given to
what they thought a perfidious Court ; and timidity,
selfishness, and dull indifference, combined to make
thousands tame and passive. But the more agitated
parts of the capital were alive with a fierce tumultuary
stir; and dark figures flitted through streets and
lanes to reach the appointed places of meeting;
while bells clanged forth from Town Hall and
steeple, as ages before they had rung out a challenge
to invading Teutonic hordes, as they had ushered in
that hour of horror and death when the kennels ran
thick with Huguenot blood. Here vehement and
gesticulating groups were seen hanging on the lips
of a fiery orator ; there conspirators sate in secluded
conclave receiving tidings from thronging messen-
gers ; in other places loud cheers greeted the gather-
ings of the mustering bands, and the quick rattle of
the drum beat a wild assembly. Meanwhile all that
was most daring had met in the sections; the form
and voice of Danton rose high and bold, though

Paris on the night of August 9, 1792.

other mob leaders had slunk off in silence; and at a given signal a body of delegates, elected by the sections with vociferous applause, made their way into the council chamber of the Commune, and, seizing on the Government of the capital, accelerated and directed the outbreak. The forces of anarchy now developed themselves; the tramp of armed columns in the streets grew dense ; the sullen clank of cannon was heard; and deep masses, headed by desperate men of hideous aspect in military garb, collected in the broad squares and ways, fringed at the edges by insurgent multitudes. Yet signs of hesitation were not wanting ; more than one tongue-valiant leader was driven on by exasperated followers threatening him with death; and the fear of Brunswick, want of mutual confidence, and the consciousness of a dangerous purpose, made many pause and turn weakly away. Hours passed before the rising attained anything like really formidable strength ; and it was daylight before the forests of pikes, here and there bristling with deadlier weapons, and skirted by yelling and enthusiastic crowds, advanced along the banks of the Seine to the thick labyrinth of enclosures and streets from which, at that time, the broad front of the Tuileries rose in antique magnificence.

Attitude of the King and the Court.

The King and the Court had during these hours been kept informed of the peril at hand. Terror and anxiety reigned in the palace ; though, at a report that the rising had failed, fine gentlemen jeered at the 'cowardice of the canaille,' and fin ladies joined in pretty disdain. Preparations were hastily made for defence; National Guards were collected from the most loyal quarters; and Pétion, Judas-like, was in attendance to screen himself and utter smooth words of hope. A handful of Nobles August 10, 1792. and their domestics, too, flocked in to strike a last blow for the throne; though the main trust of the Court lay in a few hundred Swiss, a remnant of the old Body-guard, who still lingered in the royal service. A lamentable incident, however, lessened whatever prospect of success existed. The commander-in-chief of the National Guards, a brave soldier of the name of Mandat, had prepared an able plan of resistance; and as his influence on his men was great, they might possibly not have fallen away from him. But, doubtless with the connivance of Pétion, he was lured away and murdered by the conspiring Commune; and his death left the palace without a head or leader. At the first appearance of the insurgent columns Louis went out to address the National Guards, and had he spoken and looked as became a King he might have found a way to their hearts. But the downcast bearing and hesitating gestures of the unhappy Monarch made the appeal useless; and the contempt of the crowd grew into anger when Marie Antoinette, pointing, it is said, to the few Nobles standing haughtily aloof, exclaimed, 'These are men who will show you your duty.' By this time the assailants had reached the palace, swarming round the approaches on every side;

and, far as it could gaze, the eye rested on a wild chaos of passionate wrath, of tossing steel, of menacing faces, of revengeful clamour, of hideous revelry. The weapons of the National Guards fell from their hands at the sight; and the miserable spectacle of distrust and mutiny of which so many proofs had been given was fearfully repeated at this supreme crisis. A well-meaning officer of the old Commune—Pétion had got away, his work being done—implored the King to avoid bloodshed, and to seek refuge within the Assembly, the chamber of which was a hall close by, and ill-fated Louis quietly assented. The royal family passed in sad procession along the gardens of the Tuileries, amidst the yells of ferocious mobs, baulked, for the moment, of their intended prey; and in a few minutes they were in a place of safety. The King was received with cold respect, and, indeed, many of the alarmed Assembly would have even now turned to him again if they dared; but he was soon made to feel that he was a

mere captive. A deputy having made the remark that the debates of the Assembly must be free, the royal family were huddled away into a box at the back of the reporters' gallery, and not a voice was raised of loyalty or pity. The eyes of Marie Antoinette dropped bitter tears, but the heavy features of

Louis looked dull indifference; and the chief of the illustrious race of Bourbon, in sight of the falling throne of his sires, ate, it is said with seeming content, a dish of peaches![1]

Before long the irregular sounds of disorder were lost in the din and roar of battle. The mob had forced the gates of the palace soon after the departure of the royal family; and it seemed as if the outbreak would cease, the triumph of the populace being complete. But a shot or two fired on either side caused passion to flame up more fiercely than ever; and the insurgents, headed by the men of Marseilles, made a wild dash at the inner doors of the palace. Then was seen what military worth can do against undisciplined numbers; the Swiss Guard fired and charged home, and in an instant the assailants were yelling in flight, and the refluent multitudes surged heavily backwards. At this moment, however, an order came from the unfortunate King to cease firing; and as the obedient soldiery reluctantly fell back, the revolutionary forces again pressed forward, in the exultation of unhoped for victory. A murderous and horrible scene ensued; the Swiss were hemmed in and at last overpowered; and the popular fury wreaked itself on the bodies of the dead in hideous outrage, while fiendish women danced round the mangled corpses. The palace was

The Tuileries attacked and pillaged. Massacre of the Swiss.

[1] *Souvenirs de la Terreur*, par George Duval, quoted by M. Feuillet de Conches, vol. vi. p. 285.

now stormed by the triumphant multitude; and while bands of cut-throats plied the work of murder, all that was disorderly and vile in a great city revelled in the deserted abode of royalty. In a few moments the treasures of ages were destroyed; the costly floors were strewn with the wrecks of pictures in tatters and broken statues, and troops of harlots, shrieking for the 'Austrian woman'—Court gossip had proclaimed she was as bad as themselves, and the infamous falsehood had reached the streets— were seen bedizened in the finery of the Court. Yet signs of humanity were not wanting even in these foul saturnalia of license; the ladies and women of the Court were spared, amidst shouts of 'Do not disgrace the Nation;' and a kind of principle controlled the excesses of passion, for vulgar pillage was generally forbidden, and more than one thief was caught and hanged. The lowest depths of anarchy had not yet been reached, when wickedness riots without restraint.

Reflection on the rising of August 10.

Such was the terrible outbreak of August 10, 1792, leading to the immediate overthrow of the Bourbon Monarchy. The causes of disorder which had agitated France, undermined the throne, and destroyed authority, had been made more active by various events; and foreign aggression came to give a new and extraordinary impulse to them. In the effervescence of passion which ensued, the representatives of the Nation, contending against a

Sovereign and Court believed to be false, had turned for aid to revolutionary Paris; and this power, organised by mob leaders, had overborne open and secret opposition, and displayed great and appalling strength. Authority was ere long to pass away from the classes which had so lately seized it; and the reign of license and terror was soon to prevail, with results which history will never forget. Yet many of the deeds we have briefly described were condemned by the majority of Frenchmen; and, even in Paris, the greater part of the citizens lamented the horrors of August 10. But the Constitution of 1790 gave scope to revolutionary forces; the different parties on the side of order were divided or suspicious of each other; above all, the cause of National independence and of the new interests created in 1789 seemed identified with that of the so-called patriots in circumstances which gave them extraordinary strength; and the result was that the anarchists triumphed, although really a minority in the State. It is a peculiarity, too, in the French national character, to yield easily to daring leaders; and this contributed to the fearful issue, though general causes may account for it. We shall now see how the revolutionary powers which had become ascendant went along their course, in the agony of a Nation distracted at home, and struggling to hold foreign invaders at bay.

CHAPTER V.

THE CONVENTION.

Effects of August 10.

The immediate effect of August 10 was to give the Legislative Assembly comparative freedom.[1] It might secretly dread the mobs of the capital; but it was no longer fettered by the veto and the Court; and whatever sympathy it felt for the unhappy King, it was restored, so to speak, to life by the outbreak. While the Tuileries was still in the power of the multitude, Vergniaud, the most brilliant orator of the Gironde, moved that Louis XVI. should be deposed for the present, and that a Na-

The Convention summoned.

tional Convention should be at once summoned to pronounce on the future destiny of France; and

The King and Royal Family imprisoned in the Temple.

the vote passed amids thunders of applause. Before long the ill-fated royal family was imprisoned in the Temple, an old fortress, so called from the famous

---

[1] The internal history of France, just before and during the Reign of Terror, has been described by many writers, some of them eloquent and picturesque. As an accurate and clear analysis of the events of the time, and of the working of the revolutionary institutions and press, M. Mortimer Terneaux's *Histoire de la Terreur* seems to me to deserve special notice. The notes, too, of M. Feuillet de Conches, in his sixth volume, are often valuable.

Order of that name, and was placed in the hands of
the city authorities, who claimed the charge as their
lawful right ; and the three Gironde ministers who
had been dismissed were recalled, the ministry of
justice, at the same time, being bestowed on Danton,
to please the populace. Simultaneously, energetic
attempts were made to strengthen the national de-
fences ; the camp near Paris, which had been the
subject of such fierce contention, was hastily armed ;
and commissioners were despatched to announce the
events which had taken place to the chiefs of the
armies, and to make preparations for the new
elections. Meanwhile, the usurping Commune of
Paris left nothing undone to consolidate its power
and to make the triumph of the 10th complete.
With the assent of the half-willing Assembly, the Violent
delegates of the sections annulled the existing measures of the
magistracy of the capital, and seized on its internal Commune of Paris.
police ; .and the National Guards were wholly
changed, their numbers being trebled, and their
ranks crowded with the huge bands of insurrectionary pikemen. By these means the government
of the city was secured to the demagogues and their
dependents, supported by an immense armed force ;
and though an attempt at opposition was made by
the citizens in the more wealthy sections, it was
silenced at the cry that France was in danger. By
this time the vile mob leaders who had skulked away
during the struggle of the 10th had come back to

their wonted haunts; and the Jacobins, the Cordeliers, and other places, rang with vehement ex-
hortations to the people to make use of their newly-
won liberties, and to exact a terrible vengeance 'for
the deaths of their children' from the aristocratic and
Court factions, who, with the assistance of the approaching enemy, were 'planning the extermination
of all patriots.' By the orders of the Commune,
and probably of Danton, suspected houses were, accordingly, searched, and the prisons were crowded
with hundreds of captives, who, it was openly boasted,
were detained as hostages, and marked out, in certain events, for destruction. Pétion, who, in return
for the base part he had played, had been nominally
made. mayor again, but whose influence had altogether waned, was obliged to sanction these fearful
proceedings, though he had already began to tremble
at them.

Lafayette
throws up
his com-
mand; ad-
vance of
the German
armies to
Verdun.

At this crisis a terrible incentive was suddenly
applied to this collection of passions. On receiving
the news of the deposition of the King, Lafayette
refused to obey the Assembly; and, after a fruitless
attempt to influence his troops, threw up his command, and fled across the frontier. Meanwhile, the
Austrian and Prussian armies had advanced into
the interior of France; and having rolled past the
great stronghold of Metz, were making directly, by
Verdun, for the capital, their light cavalry scouring
the plains of Champagne. All seemed lost; and

though the awestruck Assembly made passionate appeals to French patriotism, several of its leaders, especially the Gironde orators, proposed that Paris should be abandoned, and the seat of government transferred to the Loire. Danton came, however, conspicuously to the front, and declared that such cowardice was not to be thought of; and, at the same time, with a ferocious threat, certainly not understood by the other ministers, exclaimed that the real danger was from within, and that a guilty faction must be taught to tremble. Daring and unscrupulous, through less cruel than more than one of the mob leaders, he perhaps gave the signal of blood to the Commune; and, in the fury of the moment, a committee of that body proceeded to carry out the scheme of revenge which had been held out to the popular imagination. Bands of assassins were Massacre of hired to force open the prisons; and, hideous mock September. trials adding horror to the scene, their unhappy victims were ruthlessly butchered, and thrown out in heaps to crowds swarming around, amidst shouts of exulting frenzy. The execrable work of slaughter September went on for days; fear, anger, wickedness, and 1792. fiendish hate, uniting in a dreadful carnival of crime; and the complicity of the Commune is proved by the fact that it baffled several attempts to stop the triumph of blood, and its revolutionary army of National Guards was never called out to restore order, and was allowed to join or not, as it pleased, in the massacres. In this way, much that was noble and fair in a once splendid Court was ruthlessly destroyed, intermingled with numerous less known victims; and the frenzy of the murderers became so extreme that a band of State prisoners, being escorted from Orleans, was immolated with frightful cruelty at Versailles. Nor was it possible to check the devilry of passion when carnage had ceased in the emptied prisons; the form of the lovely Princess of Lamballe,[1] one of the most intimate friends of the Queen, was dragged, hideously mutilated, through the streets, and exposed to the eyes of Marie Antoinette; and ghastly processions of heads on pikes were carried through the principal streets of the capital, to strike terror into the hearts of the 'foes of the people.' At the same time pillage was let loose; the mansions of the rich and many churches were sacked; the repository of the Crown jewels was rifled; and some quarters of the city seemed like a town abandoned to a plundering enemy. Yet, as always happens, even when man appears in his most revolting aspect, faint gleams of humanity flickered here and there over these scenes of terror and dismay; noble examples of endurance and virtue were given; and many of the 'patriots' refused to

Frightful Scenes in Paris.

---

[1] Louisa of Savoy, Princess of Lamballe, was one of the purest and fairest ornaments of the Court of Versailles, and one of the few real friends of the unfortunate Queen. M. Thiers tells her death well, *Histoire de la Revolution Française,* vol. ii. p. 335, edit. 1842.

share in the spoil robbed by their baser fellows.
The victims of the butchery seem to have reached
the number of fourteen hundred persons.

Such was the ' massacre of September,' as it has
ever since been called; and, though bloodier scenes
occurred afterwards, no event perhaps in the French
Revolution was more atrocious than this sanction of
horrors by the authorities of a great capital.   Terror
and hatred account for the frightful crime, though
no excuse for it can be offered; and it is missing
the truth to ascribe it to mere party motives, or even
to any special vice in the nature of Frenchmen.
From this time the divisions between the Assembly and
the mob rulers of Paris began to grow more and more
deep; the Right, the Centre, and even the Jacobin
Mountain, concurred in the expressions of anger
and blame; and though the guilty committee of the
Commune published an eulogy on the 'justice of
the people,' hardly one even of the city demagogues
recurred with approbation to these days of blood.
In fact, the frenzy which provoked the massacre
gave way ere long to different sentiments, for a
. time at least comparatively in the ascendant, in
consequence of a sudden change of fortune.   After
the flight of Lafayette the chief command of the
French armies was given to Dumouriez; and as the
invaders began to hesitate at the critical moment
when they reached the Meuse, that general was able
to pluck safety from what .appeared peril beyond

The As-
sembly in-
dignant at
the mas-
sacre.

remedy, though he must have failed against determined foes. Drawing together all his available forces, and summoning Kellermann from Louvain to his aid, Dumouriez retreated behind the long hill range, known by the name of the Ardennes and the Argonne, which crosses Champagne just west of the Meuse; and, having seized the passes through this intricate region, he waited steadily the attack of the allies, while thousands of recruits were sent off to his camp from the capital and the adjoining Provinces. He succeeded in making a stand for a time, though driven from his positions with little difficulty; and when, at last, on September 20, the Prussians had forced his lines of defence, a misdirected manœuvre of Brunswick enabled the French, bad troops as they were, to defeat an attempt to dislodge them from the heights of Valmy.[1] This trifling advantage had wonderful results; the King of Prussia, Brunswick, and the Austrian generals, began to disagree, and to feel alarm; and the extreme wetness of an inclement season caused the invading army to perish by thousands. Before many days had passed the proud host which had advanced near Châlons was in full retreat; and France and Paris were rescued from an invasion which, if properly directed, must have crushed all resistance.

Meanwhile the National Convention had met,

Battle of Valmy. Its great results; retreat of the Prussians and Austrians.

---

[1] New and interesting details about the Battle of Valmy will be found in Feuillet de Conches, vol. vi. p. 338.

and before long was installed in the Tuileries, the Meeting of the Convention, September 22, 1792. forsaken abode of captive royalty. This Body, elected under the influence of August 10 and of foreign invasion, was more revolutionary than its immediate predecessor, but it was largely composed of the same men, and the majority were opposed to anarchy. The party of the Mountain in it, Parties in it. however, was more powerful than in the Legislative Assembly; the Plain or Centre was even more uncertain; and it was observed that several of the most distinguished deputies, who had formerly sate on the Radical Left, were now seen on the Conservative Right. The orators of the Gironde were again returned, and became the chiefs of what were called the Moderates; and the Assembly, if eager for political changes, was, on the whole, on the side of social order, though the representatives of Paris— of whom Marat, Robespierre, and Danton were the most conspicuous—were taken generally from the class of demagogues. The first measures of the Convention showed what really were its natural tendencies. A committee was appointed to enquire into France declared a Republic, Sept. 22, 1792. the charges against Louis XVI.; and Monarchy was abolished and a Republic proclaimed with hardly a single dissentient voice, the conduct of the Court during the preceding year, especially since the declaration of war, having excited general indignation and distrust. Efforts, too, were made to strengthen the armies, now pursuing the enemy

H

across the frontier; and in reply to the manifesto
of Brunswick, and the not forgotten declaration of
Pilnitz, the cause of Nations was arrayed against
that of Kings, and liberty was offered, in the name
of France, to any people who would put down its
despots. If, however, this revolutionary creed was
aggressive and even destructive abroad—and the
provocation must be borne in mind—the majority
of the Convention sincerely wished to curb anarchy
and license at home; and it viewed with alarm the
terrible events which had lately disgraced the capital.
The Moderates, led by the brilliant Gironde, denounced the atrocities of September; asserted
openly that the Commune of Paris was assuming a
power fatal to the State; and declared that Robespierre and men of his stamp were aiming at the
worst of all tyrannies.   These accusations were
generally well heard, and though fierce recriminations were uttered, though the Commune challenged
enquiry into its acts, and the clubs of the anarchists
echoed with threats, the party of mere disorder was
at first comparatively powerless in the Convention.
Savage passions, however, had been aroused; the
mobs of the capital made angry demonstrations, and
the demagogues within and outside the Assembly—
known now generally by the name of Jacobins, from
the society which was the centre of their power—began to view with deadly hatred and jealousy the
Moderates, and especially the leaders of the Gironde,

Offer of liberty to foreign nations, Nov. 19.

Dissensions renewed between the Mode- rates and Jacobins.

whose eloquence and culture provoked their resentment.

The Convention was in this disturbed state when
the report on the conduct of the King was brought
up. After an attempt on the part of the Jacobin
leaders to obtain a summary sentence of death, it
was resolved to put Louis upon his trial, and to pro-.
ceed by a regular impeachment. On December 11
the ill-fated monarch, taken from his prison to his
former palace, appeared at the bar of his republican
judges, was received in silence and with covered
heads, and answered interrogatories addressed to him
as ' Louis Capet,' though with an air of deference.
His passive constancy touched many hearts ; and such
is the sympathy that is always felt for fallen greatness when before the eye, that an immediate decision
would have perhaps saved him, though the suspicions
of the Assembly had been lately renewed by the discovery of papers of a questionable kind secreted in
an iron chest by his orders. On the 26th the advo-
cates of the King made an eloquent defence for their
discrowned client, and Louis added, in a few simple
words, that the ' blood of the 10th of August should
not be laid to his charge.' The debates in the Assembly now began, and it soon became evident that
the Jacobin faction were making the question the
means to further their objects, and to hold up their
opponents to popular hatred. They clamoured for
immediate vengeance on the tyrant, declared that

*Trial of
Louis XVI.
Dec. 11,
1792.*

H 2

the Republic could not be safe until the Court was smitten on its head, and a great example had been given to Europe, and denounced as reactionary and as concealed royalists all who resisted the demands of patriotism. These ferocious invectives were aided by the expedients so often employed with success, and the capital and its mobs were arrayed to intimidate any deputies who hesitated in the ' cause of the Nation.' The Moderates, on the other hand, were divided in mind; a majority, perhaps, condemning the King, but also wishing to spare his life; and the Gironde leaders, halting between their convictions, their feelings, their desires, and their fears, shrank from a courageous and resolute course. The result was such as usually follows when energy and will encounter indecision. On January 14, 1793, the Convention declared Louis XVI. guilty, and

Sentence of death pronounced by a majority of one.

on the following day sentence of immediate death was pronounced by a majority of one, proposals for a respite and an appeal to the people having been rejected at the critical moment. The votes had been taken after a solemn call of the deputies at a sitting protracted for days; and the spectacle of the vast dim hall, of the shadowy figures of the awestruck judges meting out the fate of their former Sovereign, and of tier upon tier of half-seen faces, looking, as in a theatre, on the drama below, and breaking out into discordant clamour, made a fearful impression on many eye-witnesses. One vote excited a sensation of disgust even among the most ruthless chiefs of the Mountain, though it was remarked that many of the abandoned women who crowded the galleries shrieked approbation. The Duke of Orleans, whose Jacobin professions had caused him to be returned for Paris, with a voice in which effrontery mingled with terror, pronounced for the immediate execution of his kinsman.

The minister of justice—Danton had resigned— announced on the 20th the sentence to the King. The captive received the message calmly, asked for three days to get ready to die (a request, however, at once refused), and prayed that he might see his family and have a confessor. A few hours afterwards the doors of his room were opened by the officers of the Commune, who stood looking on without saying a word; and the Queen, Madame Elizabeth, and the two royal children, were locked in the arms of the doomed monarch. Why raise the veil on the agony of that scene; why note too curiously the mute resignation, the passionate tears, the heart-wrung grief, of that tragic and woeful parting? Early next morning, after a tranquil night, Louis rose and gave his single attendant a wedding-ring as a token for his wife. He had promised to see his family again, but he wished to spare them the pangs of the interview. Soon afterwards he received the sacrament from the Abbé Edgeworth, a non-juring priest, who did his holy office at the peril of his life; and he

Execution of Louis XVI. Jan. 21, 1793.

remained for some time in fervent prayer, undisturbed by the sounds that rolled around the prison. At about eight the municipal officers announced curtly that the hour had come; and the King obeyed, after a few words of request that care would be taken of a will that he had made, and that a sum of money should be repaid to his counsel. He then quietly stepped into a carriage drawn up in the midst of a dense mass of bayonets, and with his faithful confessor by his side, repeated the solemn prayers for the dying, apparently unconscious of surrounding objects. The melancholy procession threaded its way through long-drawn lines of National Guards ranged on either side of the streets; and though a few sounds of anger or compassion were heard, the bystanders were rare, and for the most part silent, and shops were shut and windows closed along the course of that sad journey. For the moment pity and fear were in the minds of men ; and, in the presence of the terrible fate about to reach the descendant of a hundred Kings, even revolutionary frenzy was hushed, and the tongues of the most reckless were dumb. At ten the carriage reached a square space in view of the high front of the Tuileries; and here, near a broken statue of Louis XV., rose the guillotine, a new instrument of death. Around were deep rows of horsemen and cannon, their sabres drawn and their matches lit; a vast multitude had collected too ; and amidst the rabble of the streets was seen the familiar face of the Duke of Orleans, come again to confess his Jacobin faith. After an ineffectual attempt to address the people, drowned by the rattle of a hundred drums, the victim was placed beneath the high-raised axe; and, as the head fell, shouts of exultation burst from the lips of the vile populace, charmed hitherto as it were by a palsying spell, and a weight seemed lifted from the breasts of all.

The execution of Louis XVI. was one of those political faults which are worse than crimes. It caused profound indignation in Europe, promoted anarchy and license in France, and enlisted universal sympathy for the discrowned martyr who had borne himself so meekly in death. Those who wield power ought not to forget that a policy of bloodshed is always dangerous; and, when an august victim is selected to fall, the reaction of sentiment is sometimes wonderful. The trial, too—a mere party struggle before a popular Assembly—was a mockery of justice; and the King was innocent of the greater part of the heinous accusations made against him. But if the question be whether Louis XVI. had kept faith with the French people, and had acted in the spirit of the institutions which he had sworn to respect and uphold, History cannot record a verdict for him; and though he deprecated foreign invasion, he encouraged and dealt with the national enemy in an audacious attack on national rights. Undoubtedly,

Reflections on this event.

Character and conduct of the King. unlike our Charles I., he was not, as it were, false on principle ; he was not able enough to show kingcraft, and in his private life he was a good man, though wanting in moral and social dignity. But he repeatedly crossed the will of his subjects in a manner that looked like studied perfidy ; and he appeared to betray the dearest interests of France at a crisis when her existence was at stake—a more fatal position than any in which Charles I. was placed by the hand of Fortune. That this tortuous conduct was due to weakness, amounting to imbecility, is no doubt true. His situation also was extremely difficult ; and if we judge of his acts merely by their moral quality, we may admit that he was continually under the influence of unwise or evil counsellors. But Frenchmen, in a moment of national peril, could not draw, or trust to, distinctions like these; and had Louis been deposed when the war broke out, or even after the flight to Varennes, they would have been fully justified in the sight of posterity.

Coalition of Europe against France. The death of the King proved the signal for a general coalition of Europe against France. Such a League, indeed, had been already gathering, for the crusade of liberty which the Convention preached had exasperated every settled government ; and the progress, besides, of the French arms had been in the highest degree alarming. After his success at Valmy Dumouriez had carried the war boldly into the Low Countries, and had won a brilliant victory at Jemmapes, and by the early spring he had overrun Belgium, had advanced to the banks of the Lower Meuse, and had made an audacious raid into Holland. Another French army had seized Savoy and Nice; and a third, under Custine, had crossed the Palatinate, had taken possession of the great fortress of Mayence, and was even threatening Germany beyond the Rhine. It is not strange that the old Powers of the Continent should have viewed these invasions with hatred and fear; for the results, though caused to a certain extent by the renewed dissensions of Austria and Prussia, were evidently in the main due to the astonishing force of the new ideas which spread with the march of the French troops, and led everywhere to popular risings; and the Autocracy and Feudalism of the eighteenth century were almost necessarily led to combine against the principles of the French Revolution, which, overflowing its natural borders, was threatening with ruin their decaying authority. No definite alliances, however, had yet been made, and all was mere hesitation and doubt, until the execution of Louis XVI. fused suddenly together these blending elements, and united the rulers of all the Continental States in what they called a holy war against regicide, undertaken in the cause of God and of order. The princes of Germany followed the example already set by her leading Sovereign; Spain joined Piedmont for some time in arms; the little governments of Italy denounced

Battle of Jemmapes and early successes of the French, Nov., Dec., 1792; Jan., Feb., 1793.

France; and even Russia and Sweden stirred in their frozen deserts against the common enemy. England, too, was swept into the general movement, for the attack on Belgium had added strength to royal and aristocratic passions, and the middle classes were shocked and disgusted at the scenes which had taken place in Paris; and, amidst the exultation of the Tory party, supported by the great Whig secession, Mr. Pitt was forced into a war which he had earnestly laboured to avert.[1]   By February and

---

[1] Having reached the second year of the war, I must refer to a few authorities on the subject. A really scientific and yet popular history of the contest in a tolerably small compass is still, perhaps, a desideratum, though an approach to such a work has been made by Colonel Hamley in his *Operations of War,* in which some of the most important campaigns, from 1796 to 1815, have been reviewed with real insight and perfect fairness. The war, however, has been illustrated, in its minutest details, in numerous elaborate Histories and Memoirs, and few subjects have been treated with equal ability. Jomini has commented fully on the Revolutionary and Napoleonic campaigns; and M. Thiers, in his *Histoire de la Revolution Française,* has described the first in his usual perspicuous style, and with less partiality than he has shown in his second great work. As an account of the memorable campaigns of Italy and Germany in 1796, and of the campaigns of 1799 and 1800, Napoleon's *Commentaries* are, in many respects, unrivalled; but the Emperor is sometimes inaccurate and unjust, though incomparable as a critic of military combinations. The *History of the Consulate and Empire,* by M. Thiers, is a magnificent monument to Napoleon as a warrior; but the narrative of his exploits and those of the Grand Army is generally onesided and flattering, and should be continually checked by those of German and English writers. The campaign of 1805 is very well analysed by Colonel von Rüstow; Thiers, Alison, and Jomini may be compared for those of 1806 and 1807; and valuable papers on the operations round Ulm, and in Poland, will be found in the *Staff College Essays* by Lieutenant Baring. For the history

March, 1793, the allied armies were all in motion ; and while France was threatened from the Alps and the Pyrenees, what seemed an overwhelming tide of invasion, extending from the Scheldt to the Rhine, rolled towards her eastern and northern frontiers.

of the war in Spain and Portugal, from 1808 to 1814, the English reader will, of course, turn to the brilliant and exhaustive work of Napier; and the campaign of 1809 in Austria is well described by Generals Pelet and Stutterheim. The Russian campaign has been admirably criticised by Clausewitz and Jomini, and delineated with more or less accuracy by Ségur and Chambray. For the great struggle of 1813 and 1814 see the work of Plotho, and the narratives of Müffling, Gneisenau, and Bulow; on the French side, besides Thiers, the Memoirs of Marshal Marmont will be found useful. As for the Waterloo campaign, the authorities are almost innumerable. Mr. Hooper's account is exceedingly able and concise, but it errs on the side of praise of Wellington. Colonel Chesney, in his *Lectures on the Campaign of* 1815, has done justice to the part played by the Prussians in deciding the issue; Clausewitz and Müffling have also brought out clearly this feature of the contest; and the treatise of General Shaw Kennedy contains many valuable remarks. On the French side the *Commentaries* of Napoleon, though very unjust to his adversaries, deserve careful study; and Jomini's Précis of the campaign of 1815 seems to me very judicious in its general conclusions. Of later French writers, Thiers and La Tour D'Auvergne should be read as apologists for Napoleon, and Charras and Quinet as professed detractors and censors; but the work of Charras, able as it is, seems to me unsound and unfair. For the all-important question of the operations of Grouchy, see the pamphlet of that general, and the clear but somewhat too sanguine observations of Marshal Gerard. In addition to these and many other works on the war, the diligent reader should continually refer to the *Correspondence* of Napoleon, the *Despatches* of the Duke of Wellington, and the admirable military works of the Archduke Charles. The *Military Souvenirs* of the Duc de Fezensac are perhaps the best extant records of the characteristics and composition of the Grand Army. The naval operations of the period are set forth in the fullest detail in James's History.

Fierce
struggle of
parties in
France.

During these events the struggle between the
parties and factions which divided France had been
growing more and more fierce. The vacillation of the
Moderates and the Gironde on the occasion of the
trial of the King had increased the power of the mob
leaders; and Robespierre, who was beginning to rise
into influence by a fanatical parade of republican
doctrines, and through a reputation of austere probity,
found many opportunities to denounce what he called
the 'royalism of the Convention.' Expressions, too, of
the Gironde orators were tortured into charges that
the whole party wished to divide France into a Fede-

The
Gironde
denounced
by the
Jacobins
and Dema-
gogues.

ration of States; and this aroused intense indignation
in Paris, more especially when it was artfully proclaimed that these fine talkers had proposed to desert
the capital, a few months before, at the approach of
danger. The Gironde recriminated by fierce invectives against the Jacobins and the Duke of Orleans,
whom they accused of secretly aspiring to the throne;
but, though in the Convention they were still supreme, the revolutionary forces acquired strength,
and they suffered from the inevitable results of new
and almost usurped authority. The strife of which
Paris was the centre appeared in a thousand forms
throughout the rest of France, and usually with the
same results; the middle classes and wealthier orders
being for the most part on the side of the Moderates;
the poor, the reckless, and the discontented, taking
part with the anarchists and growing in power. All

the mischiefs, too, which had already arisen from the influence exercised by mere demagogues over the local authorities throughout the country; from the issue of assignats, now more excessive than ever; from the decline of trade which had progressed steadily, and from the pressure of poverty continually on the increase; began to tell with extraordinary force at this juncture against the upholders of social order. The rise of prices almost inevitably led to a demand that they should be fixed by the State; and measures of communism and of a maximum rate for all the principal necessaries of life were clamoured for by the popular chiefs and by the masses who looked up to them. It was in vain that the leaders of the Convention condemned such expedients as worse than useless; it is always difficult to argue with hungry men; and when Marat, with the approbation of many in the Commune, declared 'what the poor wanted was to hang the grocers,' he found thousands to echo the frightful sentiment.

Distress and social disorders.

Meanwhile the forces of the Coalition, though feebly directed and advancing slowly, had been making alarming progress. On the Rhine Custine had been driven into Alsace, and Mayence was beseiged by the Prussians and Austrians, as a preliminary step to further movements. Before long Dumouriez lost a great battle, at Neerwinden, and fell back in disorder, through Belgium, upon the French frontier; the young recruits, who formed a

Advance of the Coalition.

Battle of Neerwinden, March 18, 1793.

part of his army, disbanding at the first reverse in thousands. The North of France was thus threatened with invasion again; and the peril was increased by a quarrel between Dumouriez and the Convention, which repeated disasters envenomed and brought to an ominous issue. Dumouriez had condemned the execution of Louis XVI. and the revolutionary address to foreign nations; he complained that Jacobin sentiments destroyed discipline, and that Belgium was pillaged by the Jacobin emissaries, who had already associated liberty and rapine; and, having been called to account for his conduct, he abandoned, like Lafayette, his command, and left his army with-

Flight of Dumouriez. He throws up his command.

out a leader. At the same time intelligence arrived of a royalist insurrection in the West; and in more than one of the cities of the South, especially where the influence of the Gironde was great, the long-standing feud between the rich and the poor broke out into open civil war, and the upper classes denounced angrily the Jacobins and the mobs of Paris. These reiterated misfortunes of course embittered the strife of parties in the Convention and outside it; and in the explosion of passion which ensued everything tended to weaken the power of the Moderates and to secure ultimate success to their

Increasing power of the Jacobins. Danton. His energy.

foes. Danton, always prominent in the hour of danger, had, at the first news of the defeats in the North, brought forward a series of revolutionary schemes; and he now insisted that the one thought of Frenchmen ought .to be to save and defend the. Republic by any expedients, however desperate. The isolation of the ministry from the Legislature, which had been continued up to this time, being obviously injurious at a great crisis, he obtained a decree by which a small cabinet, chosen within the Convention, became invested with what was practically supreme authority; and thus began the Committee of Public Safety, the most terrible dictatorship, perhaps, which modern or ancient times ever witnessed. A second committee, called that of General Security, formed at his instance, obtained the superintendence of all the higher police of the country; and he procured decrees for the arrest of all suspected persons and the establishment of an extraordinary tribunal, free from most of the safeguards and checks of procedure, to coerce and terrify what he called the factions. By these means the foundations of a formidable power were laid, which might become a tremendous despotism; and, in order to provide for the national defence, Danton urged not only that energetic efforts should be at once made to recruit the armies, but that, if necessary, the whole youth of the country should be placed at the disposal of the State. To obtain the willing support of the masses, he advocated, besides, an excessive tax on the rich, violent measures to keep up the value of assignats, and, above all, the maximum of prices, the cherished scheme of the

[marginal notes:] Formation of the Committee of Public Safety, April 6 1793.

Violent measures proposed by Danton,

Parisian demagogues. 'Blast my memory,' he exclaimed, in one of those harangues which electrified the Convention with their rude force, 'but stop at nothing to save your country.'

These impassioned appeals, in which we trace a strange mixture of true insight, of absurdity, and of mere popularity seeking, were of course supported by

and decreed by Convention.

the Jacobin leaders; and, under the pressure of danger, a great part of the policy of Danton, as we

Propositions of the Commune of Paris.

have seen, received the assent of the alarmed Convention. At the same time the Commune of Paris threw itself boldly into the general movement, and, openly asserting its independence, prepared an armed force to be sent to the frontier, and called on the other cities of France to follow its example. Meanwhile, the machinery of agitation was plied with ever-increasing energy; the populace were told that now was the time for patriots, and that whoever opposed them were the foes of France; and while the Convention despatched commissioners to visit the armies and collect recruits, the revolutionary organisation which overspread the country promoted

The party of violence generally prevails.

whatever Jacobinism wished in the name of the national independence at stake. The general result was to give overwhelming strength to the rapidly growing insurrectionary forces; and even in the Convention the violent Mountain began ere long to become ascendant, and the uncertain and feeble Plain to gravitate by degrees to the more audacious

party. This consummation was accelerated by the
Moderates, and especially by the chiefs of the
Gironde, at this great and terrible crisis. As
patriotic as their opponents at least, but fearful of
revolutionary projects, and with no hold on the
popular masses, they had supported a part of Danton's
policy; but they denounced as schemes of democratic
tyranny the extraordinary tribunal and the Committee of Public Safety; and they resisted the maximum and the tax on the rich, as projects of robbery
in the name of law, though precedents for communism were not wanting. Thus, at a moment of
extraordinary peril, they thwarted what was passion-
ately announced as necessary to the National Safety;
and they crossed what its leaders took care to pro-
claim was the declared will of the Sovereign People,
and what certainly was the desire of the mob. Nor
was even this the limit of what their adversaries
called their crimes against the State. They had obtained a commission of twelve deputies to enquire
into the arbitrary acts of the Commune; and this
body had ordered two of the worst of the demagogues to be put on their trial. They had, besides,
insisted on impeaching Marat, had proposed to break
up the Commune of Paris, and to surround the Convention with a guard from the Provinces; and one
of their members had incautiously exclaimed that 'if
a hair on the head of a deputy were touched Paris
would be blotted out of the list of cities.'

The Gironde and Moderates denounce extreme measures.

The Commission of Twelve.

I

The forces of anarchy become supreme.

Thus, at a crisis of national danger, the forces of anarchy, which had been merely held in check, and had long ceased to be under control, rose again, sustained by what seemed to be the patriotic sentiment of France; while the party of order appeared vacillating, incapable of a bold resolution, and opposed to the popular demands, and it lost

Danton tries in vain to reconcile the contending parties.

weight even in the national representation. Danton, with a singleness of purpose which marks him off, stained with blood as he was, from the worse demagogues, endeavoured to reconcile the contending factions and to unite the Jacobin and Gironde leaders; but his attempts were fruitless, for it was a death-struggle. The Marats and Robespierres

Death-struggle between the Moderates and Jacobins.

hounded on the mob against what they stigmatised as a party in league for a long time to break up the Republic, and now openly plotting against France; and all patriots were adjured to support the cause of the people and of national right. The Gironde retaliated by denouncing the assassins of September, and the fomenters of trouble; but their influence daily became more weak, and power, even in the Convention, shifted from the Moderates, while they had nothing to oppose to the Commune, the Jacobins, and the Parisian populace. In this state of things their fall was at hand, and the end was not slow to

Rising of May 31 and June 2, 1793.

arrive. An insurrection very similar to that of August 10, was planned and organised; delegates from the sections of the popular type entered the

Town Hall by a preconcerted arrangement, and suddenly usurped the powers of the Commune; and on May 31 a great armed force invaded the Convention, and obtained from the deputies a decree to extinguish the Commission of Twelve, amidst frantic shouts against 'Moderates,' 'Federalists,' 'Gironde traitors,' and 'other enemies of France.' On June 2 eighty thousand national guards hemmed in the Convention, with cannon in their front; and a demand was made by the now audacious Mountain, supported by a threatening multitude, that twenty-two of the Gironde leaders should be given up and impeached for their crimes. A few courageous men protested in vain; the Plain fell off from the losing side, and the Convention decreed what was sought from it, in a state of doubt, uncertainty, and terror. The twenty-two, with seven names added, were surrendered and placed under arrest, and, the chiefs of the Moderates being struck down, the triumph of Jacobinism was complete. Thenceforward hardly anything remained to check the forces of anarchy in their career; the Convention was to follow the impulse of the Commune, and to yield obedience to the same leaders, and the Revolution was to enter on its most appalling phase.

Fall of the Moderates.

The fall of the Moderates and the Gironde was in a great part due to the causes which had produced the previous outbreak of August 10. General alarm, the result of foreign invasion, made the

Reflections on this event.

elements of disorder and passion, already too powerful, completely ascendant ; and a sentiment that the
National cause and that of the extreme revolutionists
was one, concurred with all the many incentives
which acted on the discontented and poor to precipitate and assure the catastrophe.   As for the defeated
party, it was as attached to France and her interests
as its opponents could be ; and there is no reason to
suppose that, had it continued in power, the Republic would have succumbed to the Allies.   But
the Moderates and the Gironde were wanting in
the audacity and recklessness which almost always
obtain a mastery in violent revolutions; and their
fate illustrates a general law of History.

CHAPTER VI.

THE REIGN OF TERROR.

THE Revolution of June 2 having given the party   Triumph
of violence power, its leaders proceeded to strengthen   of the party
of violence
themselves in the position they had unscrupulously   in the Convention.
won. Commanding the Commune of Paris and most
of the sections, and at last dominant in the Convention, they held the reins of government in their
hands; and their influence was sustained throughout
France by Jacobinism, by the wants of the masses,
and, largely, by national interests and sentiments.
Whatever portion of the policy of Danton remained
incomplete was now put in force; and while efforts
were made to resist the Coalition with renewed
energy, it was sought to extend and confirm everywhere the authority of the victorious demagogues
by the devices so often tried with success. The
forces of anarchy were not, however, to triumph
without provoking a resistance anarchic as themselves, and France was for some time to be torn, in   Risings
against it
the presence of her foes, by fierce civil dissensions. in the Provinces;
Some of the Gironde leaders escaped from arrest,   civil war.

and these, with other chiefs of the party, endeavoured
to excite a general rising against what they justly
described as Jacobin tyranny. Before long symptoms
of discontent appeared even in many of the Provinces attached to the principles of the Revolution;
and, at the intelligence of the fall of the Moderates,
the angry war of class in the cities of the South
broke out into inexpressible fury. Within a month
after the struggle of June 2, a large part of Normandy was in insurrection; threatening sounds were
heard in Burgundy and Alsace, in Franche Comté,
Dauphiny, and Languedoc; and the wealthier orders
being for the moment in the ascendant, Marseilles,
Bordeaux, Toulouse, and Grenoble, stood in open
revolt against the central government, to be soon
followed by Toulon and Lyons. Meanwhile the disturbances in the West, which had been menacing
for some time, assumed suddenly immense proportions; and in Poitou, Anjou, and a part of Brittany,
thousands of armed men rose up to defy an irreli-
gious and regicide Republic, to the rallying cry of

Beginning
of the war
of La
Vendée.

'God and the King.' In these secluded and remote
districts the seigneurs had for the most part lived
on their lands, and the influence of the Church was
kindly and great; and the peasantry, accordingly, had
cared little even for the good which the Revolution
had done them. But when that event had brought
with it the spoliation of the landlords they revered,
and terrible laws against their beloved priests, they

had given proofs of angry irritation, and, at the
news of the death of Louis XVI., they had expressed
their indignation in passionate risings. The severe
means by which the government forced their sons to
fight for a detested cause filled up the measure of
their discontent, and by the middle of 1793 they
had formed great insurrectionary bands, which were
to prove far from contemptible foes. Such was the
beginning of the celebrated war of La Vendée—so
called from a Department of that name—one of
the darkest episodes in the revolutionary drama.

These perils, though added to those of foreign
war, did not, however, prostrate the energies of the
men who were now supreme in France. Vile and
worthless as many of these leaders were, some were
not wanting in daring and constancy; and Danton
urged the Governing Powers to redoubled efforts. A
levy of three hundred thousand men which had been
voted was ordered to the frontier; and while preparations were made to enforce the decree for what
was called 'levée en masse,' the Jacobin leaders
turned against their domestic foes. The extraordinary force which had been set on foot in Paris,
and which received the name of the revolutionary
army, was marched against the insurgent districts,
with as many National Guards as could be spared;
the cities in revolt were summoned to yield; and
emissaries were despatched to stir up the masses
against the 'enemies of France and the allies of the

Energetic
measures of
Danton and
the leaders
in power.

The levée
en masse.

The Constitution of 1793.

The maximum. The Revolutionary Tribunal.
stranger.' Meanwhile a Constitution of the most democratic type was offered as a rallying-point to the People ; good patriots were commanded to form everywhere revolutionary committees to support their leaders ; the maximum and the tax on the rich were announced as assuring universal comfort to the poor ; and the extraordinary, now styled the Revolutionary Tribunal, began to send daily its victims to the guillotine, while the prisons were filled with suspected persons. These measures were attended with astonishing success, though, but for deeper causes, they would have certainly failed. Lyons and Toulon, indeed, long remained in arms ; and the rising of La Vendée, sustained by a principle, and at first encountered only by levies of recruits, became in the highest degree formidable.

The risings in part of France quelled.
But the insurrection in the North was quickly dissipated ; most of the Provinces became soon quiescent ; and before long nearly all the Southern cities were overawed or tamed into submission. This rapid collapse, as we have said, was due to causes beyond the mere acts of the Jacobins, though these unquestionably were not fruitless. The authority of the central government was immense.; and when Jacobinism had laid hold of the capital it quickly triumphed in other parts of the country, already largely controlled by it. Besides, the feud which divided France was generally one between the needy and the well-off; and in the existing state of all in-

stitutions, the needy were certain to prevail, even apart from the tremendous stimulants supplied lavishly to the wildest passions. Moreover, strongest motive of all, the cause continued in full force which had made Jacobinism succeed at first; and it seemed treason to the State, and fatal to France, and to all that had been done since 1789, to oppose Danton and his supporters when they hurled defiance against the foreign invader.

From these causes the civil war in which France for a moment appeared engulphed was soon confined to a few narrowing centres. What, in the meantime, had been the achievements of the mighty Coalition of banded Europe? Success, that might have been great, was attained on the Alpine and Pyrenean frontiers; and had the Piedmontese and Spaniards been well led they could have overrun Provence and Rousillon, and made the insurrection of the South fatal. But here, as elsewhere, the Allies did little; and, though defeated in almost every encounter, the republican levies held their ground against enemies who nowhere advanced. It was, however, in the North and the North-east that the real prize of victory was placed; and no doubt can exist that had unanimity in the councils of the Coalition prevailed, or had a great commander been in its camp, Paris might have been captured without difficulty, and the Revolution been summarily put down. But the Austrians, the Prussians, and the English, were

Feebleness of the Coalition.

divided in mind; they had no General capable of rising above the most ordinary routine of war; and the result was that the allied armies advanced tardily on an immense front, each leader thinking of his own plans only, and no one venturing to press forward boldly, or to pass the fortresses on the hostile frontiers, though obstacles like these could be of little use without the aid of powerful forces in the field. In this manner half the summer was lost in besieging Mayence, Valenciennes, and Condé; and when, after the fall of these places, an attempt was made to invade Picardy, dissensions between the Allies broke out, and the British contingent was detached to besiege Dunkirk, while the Austrians lingered in French Flanders, intent on enlarging by conquest Belgium, at that period an Austrian Province. Time was thus gained for the French armies, which, though they had made an honourable resistance, had been obliged to fall back at all points, and were in no condition to oppose their enemy; and the French army in the North, though driven nearly to the Somme, within a few marches of the capital, was allowed an opportunity to recruit its strength, and was not, as it might have been easily, destroyed. A part of the hastily raised levies was now incorporated in its ranks; and as these were largely composed of seasoned men from the old army of the Bourbon Monarchy, and from the volunteers of Valmy and Jemmapes, a respectable force was before

Waste of time, in action and dissensions.

long mustered. At the peremptory command of the Jacobin Government, this was at once directed against the invaders, who did not know what an invasion meant. The Duke of York, assailed with vigour and skill, was compelled to raise the siege of Dunkirk ; and, to the astonishment of Europe, the divided forces of the halting and irresolute Coalition began to recede before the enemies, who saw victory yielded to them, and who, feeble soldiers as they often were, were nevertheless fired by ardent patriotism.

Sept. 8, 1793.

As the autumn closed the trembling balance of fortune inclined decidedly on the side of the Republic. The French recruits, hurried to the frontiers in masses, became gradually better soldiers, under the influence of increasing success. Carnot,[1] a man of great but overrated powers, took the general direction of military affairs ; and though his strategy was not sound, it was much better than the imbecility of his foes. At the same time, the Generals of the fallen Monarchy having disappeared, or, for the most part,

The Republic successful at home and abroad.

Carnot. Hoche.

---

[1] Lazare Nicolas Marguerite Carnot was born in 1753, and brought up to be an engineer. He distinguished himself in his profession, and at the crisis of 1793 was made Minister of War, and became one of the Members of the Committee of Public Safety. His energy was above praise ; but though it has been said of him that ‘ he organised victory,’ his military schemes were often unsound. He was, however, the only member of the Committee whose hands were, in some degree, free from the stain of blood. In after life he was exiled, opposed the Empire, supported it in the hour of danger, distinguished himself for his defence of Antwerp, and served as Minister of the Interior during the Hundred Days. He outlived Waterloo several years.

failed, brilliant names began to emerge from the ranks, and to lead the suddenly raised armies; and though worthless selections were not seldom made, more than one private and sergeant gave proof of capacity of no common order. Terror certainly added strength to patriotism, for thousands were driven to the camp by force, and death was the usual penalty of a defeated chief; but it was not the less a great national movement, and high honour is justly due to a people which, in a situation that might have seemed hopeless, made such heroic and noble efforts, even though it triumphed through the weakness of its foes. Owing to a happy inspiration of Carnot, a detachment was rapidly marched from the Rhine, where the Prussians remained in complete inaction; and with this reinforcement Jourdan gained a victory at Wattignies over the Austrians, and opened the way into the Low Countries. At the close of the year the youthful Hoche, once a corporal, but a man of genius, who had given studious hours to the theory of war, divided Brunswick from the Austrian Würmser by a daring and able march through the Vosges; and the baffled Allies were driven out of Alsace, the borders of which they had just invaded. By these operations the great Northern frontier, the really vulnerable part of France, was almost freed from the invaders' presence; and, though less was achieved on the Southern frontier, the enemies of the Republic

Battle of Wattignies, October 16, 1793.

began to lose courage. Meanwhile Lyons had fallen Fall of Lyons,
after a terrible siege ; and though the struggle in October 9, 1793.
La Vendée was not over, the cause of the royalists
was rapidly declining. On this theatre the Catholic army, as it called itself, had won a series of
triumphs ; and the peasant bands, commanded by
their seigneurs, · and largely composed of excellent
marksmen, proved more than a match, in an intricate
country, for revolutionary recruits and generals
chosen from the noisest spouters of the Commune of
Paris. At last, however, the garrison of Mayence,
and a real commander, Kleber, appeared on the
scene; and science and skill inevitably prevailed,
though the contest was protracted and desperate.
After the loss of a great battle at Cholet, in Poitou, Great defeat of
the Vendeans were driven north of the Loire ; and the Ven-deans at
before long the remains of their forces were well- Savenay,
nigh annihilated on the field of Savenay. The in- Dec. 23, 1793.
surrection had been so formidable that a few months
before they might not improbably have marched to
Paris and seized the capital.

Towards the end of December a memorable in- Siege and fall of
cident brought the eventful struggle of the year to Toulon, Dec. 19,
a close. Toulon had, as we have seen, revolted ; 1793.
and the citizens of the upper and middle classes
had unhappily called in the allies to aid them. An
English and Spanish fleet, accordingly, had taken
possession of the port and the arsenal ; and though
the town had been partly invested, the siege, con-

ducted by incapable men, made no progress for
several months. A plan of attack was at last despatched from Paris; but at a council of war a youthful artillery officer, as yet only in a subordinate rank,
observed that regular approaches were useless, and
that if a point were taken which commanded the
roadstead, the allied fleets would certainly make off,
and an immediate surrender be the consequence.

First appearance on the scene of Napoleon Bonaparte. Putting his finger on a promontory marked on a
map, he said decidedly, ' There is the key of Toulon;'
and his audience was so struck with the evident
truth that it ventured to neglect the government
order, and allowed its young adviser to work out his
project. After a sharp engagement the point was
occupied; and the French batteries had no sooner
crowned the heights than the allied squadrons made
haste to depart, and Toulon was in a few days in the
hands of the victors. This remarkable exploit was
the prelude to a career at which the world grew pale.
The young artillery officer was Napoleon Bonaparte,[1]
the mightiest product of the French Revolution.

The Reign of Terror. Meanwhile, under the double influence of foreign
war and peril at home, the anarchic forces which
had become ascendant had consolidated themselves
into a fearful tyranny, and the period known by the
ominous name of the Reign of Terror had opened on

---

[1] The biographies of Napoleon are innumerable. A very able,
but unfavourable, account of his early life and career, will be found
in M. Lanfrey's *Histoire de Napoleon I.*

France. The Convention, after the fall of the
Moderates, became a mere instrument of the Jacobin
chiefs ; seventy-three of its members were arrested
for a secret protest against the 2nd of June ; the
Right, baffled and suspected, cease to struggle ; and
the Plain registered the decrees of the Mountain,
itself bowing in passive subjection to the terrible
Committee of Public Security.   That Body, formed
of extreme Jacobins, drew to itself all the powers in
the State ; and, with the national representation in
its hands, and controlling and directing the Com-
mitte of General Security, the Commune, the clubs,
and the revolutionary committees which spread
their network over the country, it exercised an
appalling despotism.   Under this extraordinary
scheme of government the whole resources of France
were grasped by a knot of audacious and desperate
men ; and the most violent effort ever beheld was
made, not only to crush the enemies of the State,
but to turn society upside down, to subvert all its
ordinary relations, to change the usages, the habits,
and the faith of the nation, and to overbear opposition by sheer terror.   The *levée en masse* was rigor-
ously enforced, and every man, woman, and child in
France was ordered to aid in the national defence,
while the whole products of the country were declared to be ' in requisition ' for the use of the Republic.   The lands and goods of *émigrés* and of
prisoners of State were confiscated by a summary

The Convention, a mere instrument of the Jacobin leaders.

The Committee of Public Safety all powerful.

Its tyranny and terrible expedients.

process; and decress of the most ferocious kind were levelled against the revolted cities. Measures of frightful severity were taken against the unhappy class now known by the name of 'suspects;' an ampler and freer sweep was given to the guillotine; the Revolutionary Tribunal was made permanent; and expedients, such as were never heard of, were adopted to keep up the failing assignats, while the maximum was extended to almost all commodities; attempts were made to regulate the consumption of the nation; the National Debt was what was called republicanised, that is, to a great extent, wiped out; and the systematic plunder of the rich became a regular device of government. Death was the normal penalty for the slightest complaint against this widespreading scheme of oppression; nay, even for lukewarmness or 'want of *civism*;' and a failure in the field was usually followed by a mandate from the Republican commissioners, who attended the armies,

Wild social changes. for speedy execution. [At the same time a complete revolution was made in dress, manners, and even modes of speech; the very forms of language were violently changed; the Calendar and the whole system of measures were transformed; and though the Committee of Public Safety did not yet publicly proscribe the Christian faith, they regarded with aversion and distrust priests of all kinds, non-juring

Atheism declared truth by or otherwise; atheism was proclaimed truth by the Commune of Paris—an example imitated by other cities; and the churches were everywhere handed the Com-
mune of
Paris. over to the municipalities and local authorities, to be shut up or destroyed at their pleasure.]

The scenes witnessed during this strange period Appear-
ance of
Paris
during the
Reign of
Terror. of tyranny in union with popular license brought out human nature in its most stern, most terrible, and most ludicrous aspects. Paris seemed turned into a vast camp, hundreds of smithies and forges filling the squares, for the manufacture of arms and cannon, the streets barricaded and patrolled by pike-men, the houses lettered with the names of their in-mates; while young men were hastily drilled in thousands, old men and women were told off in bands to 'excite patriots to revolutionary work,' and chil-dren scraped lint and made bandages, amidst mob oratory and wild airs of music. Long lines of faces were seen at the bread-shops, waiting for the supplies fixed and priced by the State; and government emissaries filled the establishments once dedicated to the splendour of Versailles, to enforce the maximum for 'good citizens.' Informers crowded the banks and the Exchange, to mark down anyone who dared to cheapen assignats; and these pieces of paper, con-verted literally into tickets of plunder by the rule of terror, paid debts, and served to transfer commodi-ties, at their nominal value, to some extent at least, although rapidly becoming worthless. Meanwhile commissioners, 'in the name of the Republic,' seized and piled in storehouses whatever was needed 'for the armies of the patriotic poor ;' and it fared badly with those who dared to look clean, to dress well, to wear a watch or a trinket; for if not hurried off to prison as 'suspects,' they were freely relieved of superfluous luxuries. Similar sights were seen in other great cities; and the chief roads swarmed with

The levies hurried to the frontier.

masses of recruits rolling to the frontier, in varying moods of fear, regret, and fiery exultation, while the crops, the stock, aud the horses of the peasant, were numbered or taken by flocks of officials, their owners sometimes looking on in despair, but more often exclaiming that, after all, ' France and the Revolution

Appearance of the Convention.

must be saved.' The hall of the Convention, at the same time, echoed with strange debates, and still stranger reports, in which a jargon of pagan antiquity mingled with vulgar ribaldry and the slang of fish-wives, vociferously applauded by overflowing galleries; and the same eloquence was heard in all other assemblies, but usually at a still lower level. The prisons, meanwhile, grew more and more full with ever-increasing lists of 'suspects;' and even the fearful means by which they were cleared could not keep down the vast tale of victims. Nine or ten men, of whom the most conspicuous were Robespierre, St. Just, Couthon, Collot D'Herbois, Billaud Varennes, and Barère,[1] sate in a small closet in what

---

[1] The leading Terrorists, as was the case too during the reign of the Commune of 1871, were for the most part men of low origin, and broken fortunes. St. Just, a fanatic like Robespierre, was an un-

had been the Tuileries palace, directing the immense organisation of force by which France was moved and controlled.

In this ecstacy of revolutionary passions and over-turn of social relations, whatever was most violent was sure to prevail; the long-standing vices and ills of the State provoked a retribution worse than themselves; and the popular frenzy displayed itself in excesses of license which knew no limits. As has been observed in similar movements, the very signification of words was changed; and pitilessness became Republican virtue, moderation culpable treason to France, inexorable severity patriotic devotion, atrocious cruelty irregular justice. Then, too, were seen in their worst aspect the ill-will and hatred engendered by the differences of class in the old Monarchy; to be an 'aristocrat' was in itself a crime: the few Nobles and prelates who lingered in France were either condemned, or usually shrank out of sight; and the popular exasperation rose even fiercer against the professional and trading orders, the lawyers, the merchants, the employers of labour, the dependents of the Court, the old servants of the State, the horror-stricken reformers of 1789. In the jealousy against all eminence which prevailed, even

Dissolution of society and of morality.

Cruelty and suspicions of populace

known student. Couthon was a cripple. Collot d'Herbois had been an actor hissed off the stage. Billaud Varennes had been expelled his father's house, and had been a tailor, an actor, and a dependent of the Jesuits. Barère, the 'Anacreon of the Guillotine,' began life an obscure man of letters, and ended it a traitor and a spy.

the aristocracy of intellect was denounced; men of letters and science were largely proscribed, and art and learning were either degraded to mean uses or

were declared dangerous. Licentiousness, too, broke through all bounds in the general collapse of old social restraints; the increase of concubinage, of divorces, and of illegitimate births, alarmed even Jacobin politicians; and the vices of the great were wildly imitated, with reckless indecency, by the multitude. Perhaps, however, the most striking sign of the times was the manner in which religion was treated. Christianity, we have said, was not yet disavowed by the State; but in hundreds of places the churches were stripped of their ornaments by exulting mobs; and the profession of atheism by the Commune of Paris was celebrated by a ceremony in which a painted harlot was installed in the aisles of

Notre-Dame, and hailed as the Goddess of Reason; while festivals, pagan in their character, commemorated the prolific powers of the seasons. Too often, besides, priests were found who denied the faith of which they were living witnesses; the mysteries of Christianity were profaned by one perjured bishop in a revolting parody; and much that was foul and hideous came out from under that august Church which had been long tainted by sin and corruption. Yet these blasphemies were by no means general; and thousands of the clergy, pursued as they were by Jacobin suspicion, continued to perform their holy

offices to reverent congregations, who still adhered
to the creed of their fathers. Nor was all evil even
in this fearful season of national trial and social subversion; noble instances of fidelity and virtue were
seen, apart from the patriotism which inspired
Frenchmen ; and a kind of distempered public spirit
may be traced in the scheme of Jacobin policy, extravagant and iniquitous as it was.

The march, however, of the Reign of Terror has   Scenes in
the prisons.
yet to be viewed in its most tragic aspect. The
prisons, we have said, were thronged with victims
whom ferocious laws, or ruthless suspicion, or private
malice, sent to their precincts; and, in Paris alone,
the number of captives was usually from five to six
thousand persons. In these dark and terrible abodes
were packed in masses— without regard to distinctions
of rank, of age, of sex—the noble, the beautiful, the
highly refined, with the vile, the worthless, and the
merely criminal; the seigneur, the court dame, the
man of taste, chiefs of the National and Legislative
Assemblies, unfortunate generals, discarded magistrates, priests, merchants, and caterers for the luxury
of Versailles, confusedly mingled with forgers and
thieves, and the most degraded refuse of the streets.
Eye-witnesses have left vivid descriptions of what
occurred in these frightful Assemblies; how human
nature became desperate, or reckless, or callous, or
even mirthful, under the influence of continued
suffering ; how social differences were jealously pre-

served or vanished in the presence of common peril;
and how virtue asserted its natural authority in the
disappearance of conventional forms; and the depraved treated the good with respect, while they
persecuted the vicious, whatever their station. A

collection of prisoners was almost daily consigned to
the Revolutionary Tribunal; and though that murderous court was not yet at its worst, its ordinary
process was swift and fearful. The condemned were
hurried off to the guillotine, and, in the presence of
revelling crowds of the most ruthless and cruel populace, were usually slaughtered in batches at a time,
amidst clamorous shouts of ' Long live the Republic.'
So perished, with numbers of less known victims,
not a few of the most illustrious names of France,
surviving ornaments of the old order of things, brilliant popular leaders of a few years before. Several
of the Gironde deputies had died miserably in the
rising of June 2, and Pétion, among them, had met
a fate which History cannot call undeserved; but
most of the arrested twenty-two were sacrificed, with
Vergniaud, the most eloquent Frenchman of his time.
Such, too, was the doom of the once famous Bailly,
of the high-souled and chivalrous Barnave, of the
infamous and recreant Duke of Orleans,[1] of Custine,

---

[1] The death of this disgrace to his name is thus described by
Mr. Carlyle:—' Philippe's eyes flashed hell-fire for an instant; but
the next it was gone, and he sate impassive, Brummellean polite.
On the scaffold Samson was for drawing off his boots. " Tush,"

and other distinguished officers, of Malesherbes, the great advocate of Louis XVI. But why dwell further on the appalling record? Two deaths, however, strikingly showed what was most noble in the social life which the Reign of Terror endeavoured to destroy. The fair and saintly Madame Elizabeth [1] drew tears even from Jacobin eyes, as, piety struggling with maiden shame, she bowed her head meekly to the fatal axe; the sterner but heroic Madame Roland,[2]

said Philippe, "they will come better off *after*. Let us have done; dépêchez-vous." '

[1] 'The only emotion she showed,' says an eye-witness, 'was when the executioner approached her to remove her shawl. "For Heaven's sake, sir," she exclaimed, "spare me the exposure!"'—*Six Jours au Temple*, p. 75; F. de Conches, vi. p. 556.

$$\text{ἡ δὲ καὶ θνῄσκουσ' ὁμῶς}$$
$$\text{πολλὴν πρόνοιαν εἶχεν εὐσχήμως πεσεῖν,}$$
$$\text{κρύπτουσ' ἃ κρύπτειν ὄμματ' ἀρσένων χρεών.}$$

[2] Marion Jeanne Phlipon, Madame Roland, one of the most celebrated characters of the Revolution, was born in 1754. She was the daughter of an engraver, and her Memoirs show how, even in early life, she resented the distinctions between the Noblesse and Bourgeoise. In her teens she gave proof of the energy, the fervour, and the sentimentalism of her mature years; but she grew up a sceptic, fed on the false literature and philosophy of the day. In 1780 she married M. Roland, then an inspector of manufactures at Rouen, and, soon after the beginning of the Revolution, repaired with her husband to Paris. There she became Queen of the Gironde party, and, when M. Roland was made Minister, his chief adviser. Her Memoirs illustrate the enthusiasm, the genius, and the unpractical conduct of the Gironde orators, and throw a vivid light on most of the events which led to the Reign of Terror. She was involved in the proscription of the Gironde, and perished on the scaffold in 1793. Her death had something grand and yet theatrical about it; and though her character was noble, it was hardly womanly, and was too artificial to charm.

the celebrated wife of a minister of that name, went
with a smile on her lip to the scaffold, exclaiming, 'Liberty! oh what crimes are done in thy
name!'

Trial and
execution
of Marie
Antoinette,
October
14-16,
1793.

On October 14, 1793, Marie Antoinette was
brought before the fatal tribunal. Her appearance
filled for an instant with pity the hearts even of her
hardened judges, and of the barbarous audience
which thronged the hall. The hair of the Queen
had turned white; grief had furrowed prematurely
her noble countenance; she was arrayed in a coarse,
miserable garb, which hung loose on her still stately
form; and in the light which drew out her figure
from the dim benches and galleries around, she looked
a wreck of oppressed majesty. How different from
that vision of youth and grace that had once flitted
along the terraces of Versailles; how changed from
that princely yet winning presence so often greeted
by applauding multitudes, so long the centre of the
homage of chivalry! But the sentiment of compassion passed away, for Marie Antoinette was
abhorred and feared, and the mockery of a trial
quickly went on. More than one personage of the
late Court, willing to barter honour for the chance
of safety, bore witness against the doomed captive;
and a nameless and execrable charge was made which
received an answer of such pathetic truth that even
the foul-hearted accuser was silenced. Sentence was,
of course, before long pronounced, and on October

16 the victim was led to the guillotine. Forms of
decency had long ago disappeared; and Marie
Antoinette was drawn to the place of execution, exposed to the insolent gaze of the populace, in a
common cart, with her arms bound, in a prison
dress, like the vilest criminal. The calm dignity,
however, which had more than once abashed her
judges a few hours before, did not desert the Queen
in her last moments; and it was observed that several
of the woman fiends who crowded round to yell as she
passed shrank from her steady and serene gaze. On
the fatal journey she seemed perfectly composed,
except when, in the words of an eye-witness, ' her face
gave signs of lively emotion ' at the sight of what
had been once the Tuileries; and she encountered
death without display or flinching. Her end was
noble, and the foul slanders which gathered against
her pure life were falsehoods; and we need not enquire what, in her case, was the iniquity of the
Revolutionary Tribunal. But it is not the less true
that Marie Antoinette, like Louis XVI., had wronged
France; and the wrong she had done was the more
grievous in that she was a chief counsellor of her
imbecile husband, and he was mere clay in her proud
hands. Still, in judging her conduct, the associations   Her character and
of her life and of her situation must be fairly weighed;   conduct.
and History, as it marks that stately figure, tossed,
feebly resisting, over the abyss, may well muse on
the tyranny of circumstance, and echo the truth that

the Tower of Siloam may fall on those not the most guilty.

Divisions among the Jacobin rulers. For several months few changes were made in this system of widespread tyranny ; and the men who had seized on power in France forgot or sunk their differences under the stress of danger. When, however, the Republic emerged from its first trials, divisions sprung up among the Jacobin chiefs; and three parties gradually developed themselves, representing the conflicting views of their leaders. Dan-

Three factions form themselves. ton, who, even as early as July, had quitted the Committee of Public Safety, inclined before long to the side of clemency; and his wishes were seconded by a large following, who looked up to him as the champion of the revolution. These men, turbulent and savage as they were, had nevertheless human sympathies and feelings; they were not maniacs of fanatical principles, and they aimed rather at enjoyment and influence than at any fixed Republican ideal; and though, like Danton, they were morally corrupt, they had desired to spare the Gironde victims, and began to condemn the excesses of the Reign of Terror. The second faction, led by a wretch called Hébert,[1] was composed of the extreme anarchists of the Commune of Paris, who had preached

---

[1] Jacques Réné Hébert, born in 1755, was a footman and a box-keeper at a theatre, and had lost both places for dishonesty. When the Revolution broke out he became Editor of the *Père Duchesne*, the most indecent and ribald print probably that has ever seen the

atheism, and given the freest rein to license ; and
the political object of these miscreants was to make
the capital supreme in the State, and to secure independence to the great cities, while their social creed
was mere sensual indulgence.   The third party was
led by Robespierre, and by degrees it became the   Growing
strongest, for the reputation of that singular being   ascendency
of Robes-
had gained for him a great moral ascendency ; and   pierre.
the views he professed with a parade of virtue fell in
largely with the popular sentiment, always gratified
when its worst aspirations are flattered in the name
of the public good.   The hope of Robespierre and
his immediate followers was to set up a Republic in
accordance with the wild and mischievous notions of
Rousseau ; and as this end could not be approached
without carrying out relentlessly the system of Terror,
they condemned what they called the moderation of
Danton, while they abhorred, as opposed to their
theories, the godless licentiousness of the Commune
demagogues.   Robespierre, though possibly not cruel
by nature, was, like all men of his type, pitiless when
ruled by the ideas on which he had brooded ; and
this was the character of one or two of his chief sub-
ordinates, though the great mass of the party were
mere Jacobins, yielding to that impulse which always

light, though an imitation of it appeared during the Jacobin
saturnalia of 1871. This miscreant became one of the chief officers
of the Commune of Paris, and it was he who made the unnatural
and foul charge against the Queen alluded to above.

secures authority for a resolute faith, sustained by real or seeming probity.

He becomes supreme in the State. Before 1793 had closed, the ascendency of Robespierre was complete. He was the especial favourite of the Jacobin Club; his influence in the Convention was supreme, and he was the dictator of the Committee of Public Safety. The dissensions between the hostile parties soon broke out into open discord, and personal antipathies deepened the feud. With the system of government which prevailed, the possessors of power could easily destroy their rivals; and Robespierre and his satellites turned without scruple the tremendous machinery in their grasp against their adversaries on either side. Under the pretence of conspiracies, of which proofs were always forthcoming in an atmosphere of preternatural suspicion and passion, Hébert and the leaders of the

Destruction of Hébert and the leaders of the Commune, of Danton and his followers, March 24, April 3, 1794. Commune were first swept away, and with their fall that famous organisation which had been a mainspring of the Revolution, and had made Paris dominant in the State, lost a great deal of its immense influence. The turn of Danton and his chief friends came next; and though the struggle was perilous and long, they too passed before the Revolutionary Tribunal, and were immolated by means of a special decree obtained from the overawed Convention. With them perished what may be described as the Moderates of the Reign of Terror, and compassion must be felt for the fate of their leader. Danton was a man of great natural powers ; courageous, resolute, Character of Danton. with a genius for command, with an eloquence, rude, but of extraordinary force ; and if the blood of Sep-. tember be on his head—and he often played the demagogue for his own ends—he had, nevertheless, a patriotic heart; he is entitled to any merit which belongs to the Jacobin scheme of national defence ; and it is to his lasting honour that he risked and lost his life in the sacred cause of humanity. After his death Robespierre and his creatures became the absolute masters of France, and they lost no time in strengthening their sway. The authority of the Dictatorship of Robespierre. Committee of Public Safety was made more complete than it had ever been; and in order to keep down the Commune of Paris, the revolutionary army was disbanded, and the democracy of the sections was in a great measure controlled, while the chief magistrates were chosen from dependents of Robespierre. At His measures to secure his power. the same time clubs and popular societies, with the one exception of the trusty Jacobins, were suppressed by a summary mandate; and, as if to show what a Republic of virtue was to be, atheism was pronounced ' an aristocratic falsehood,' the worship of ' the Supreme ' was declared the national faith, and Chris- The worship of the Supreme. tianity was proclaimed a base superstition, and its ministers criminal dupes or impostors.

And now, mastered by Robespierre, the Reign of Terror at its height. Terror quickened its march, and grew more fearful in its murderous activity. A merciless fanatic

swayed the small oligarchy of which the powers had
been just increased; and, as if to prove what
Jacobin 'freedom' was, the worst deeds of which
the old Monarchy had been guilty in the course
of ages were infinitely surpassed in a few months,
under a form of government in many respects
similiar. A decree was wrung from the oppressed
Convention by which the Revolutionary Tribunal
was set free from all checks, and 'moral conviction'
was made sufficient proof of crime; and the energy
of that instrument of slaughter became suddenly
more than ever appalling. Prisoners were tried by
forties and fifties at a time, and sent to their doom
with summary glee at a nod or a wink of infamous
accusers; and—a fitting emblem of the revolting
scene—the guillotine appeared in the place of judgment. 'Suspects' were crammed, literally in thousands, in dens, in which vile informers glided about,
making sure of the means to do them to death; and
when other charges could not be made 'conspiracies
in the prisons' were feigned to serve the purpose.
The dread and agony which had taken possession of
all within the possible reach of this frightful
tyranny proved often too much for nature to endure; and suicides and madness awfully increased,
while Paris bore the look of a city abandoned to a
mere multitude of reckless barbarians, what was
orderly and decent having cowered out of sight.
Meanwhile, the system of spoliation inaugurated

*Frightful
state of
Paris.*

by the maximum and forced assignats was carried
on more stringently than ever; and as authority
had become fully concentrated, devices of escape
grew more difficult.   At the same time the most
atrocious vengeance ever witnessed perhaps in
western Europe was wreaked on the hapless revolted
cities.   Attempts were made to raze Lyons and
Toulon to the earth ; and ' floods of death,' as it was
said, ' swept away traitors and moderates' in these
devoted places.   Similar horrors were seen at Bordeaux, Arras, and Marseilles ; and for miles below
Nantes the Loire rolled to the sea hundreds of
corpses twisted in ghastly embraces, the victims of
what, with hellish mirth, were designated as ' republican marriages,' having been tied together, and,
crowded in barges, deliberately scuttled and then
sent adrift.   Simultaneously La Vendée, still in
part insurgent, was traversed throughout by 'infernal
columns;' and, notwithstanding a manly protest of
Kleber, who foresaw the inevitable results, these
bands everywhere marked their advance by murder,
pillage, and widespread havoc.   Commissioners, despatched with ' full powers' from the capital, urged
the populace, wherever they could, to these crimes;
and Robespierre was the sovereign head and absolute
lord of this system of blood.   If in the *chambres
ardentes* of the Bourbon monarchy, in the frequent
oppression of the old Parliaments, in the horrors of
the Bastille and other State prisons, in the massacres

Massacres in the Provinces.

of St. Bartholomew and at La Rochelle, in the centralised, cruel and suspicious governments of more than one of the Kings of France, we see a faint foreshadowing of this order of things, tyranny so rapid and deadly had never before been witnessed ; and few probably will think that the execrable character of the last and worst phase of the Reign of Terror was mitigated by blasphemous festivals to ' the Supreme,' or even by empty and illusory projects to 'abolish poverty' and other social evils.

Such was the fulfilment of the glowing hopes which had animated France four short years before ; such was the practical issue of the philosophy which had dazzled a generation by its glittering chimeras. The land was a land of mourning and carnage ; and the Rights of Man terminated in a ruthless despotism sustained by the worst dregs of the masses. And what made this tyranny the more atrocious was that the impulse was failing which had first given the Jacobins overpowering force ; for, instead of being threatened with destruction, the Republic was entering on a career of victory. The discomfiture of 1793 had made the Allies more than ever divided ; the long-standing jealousies of Austria and Prussia were aggravated by intrigues about Poland; and

The Republic obtains fresh successes in the campaign of 1794.

when the war was renewed in the spring of 1794, the Coalition was ill-prepared to encounter a daring and resolute enemy. Meanwhile, the gigantic efforts of France had been attended with great results, and fully half a million of men stood in arms on her frontiers to confront her adversaries. The consequences were such as usually follow a struggle between discordant weakness and earnest and enthusiastic strength, though other and potent causes concurred. The new French levies, indeed, were still often defeated, even with a large advantage of numbers on their side ; and, without an admixture of trained soldiers, they still proved comparatively worthless. On the sea too, the hastily equipped fleets of the Republic met a crushing reverse ; and the great victory of June 1 gave England the first of a long series of triumphs. But numerical force, union, and patriotism told ; and they were aided by a direction at least always better than that existing in the hostile camps. The Spaniards were driven behind the Pyrenees; Savoy and Nice were brilliantly regained; and the young conqueror of Toulon, baffling the Piedmontese by one of those manœuvres which began to show his powers, beheld, Hannibal-like, from the tops of the Alps, the plains soon to be the scenes of his most splendid exploits. Meantime, after a protracted struggle, the Duke of York was beaten on the Belgian frontier ; and while Pichegru and Moreau advanced into Flanders, Carnot repeated the operation of the preceding year; and, profiting by the remissness of the enemy in the Vosges, moved a considerable force from the Meuse to the Sambre, which gave the French victory

English naval victory of June 1.

The allies defeated on all other points of the theatre. Battle of Fleurus, June 26, 1794.

L

on the plains of Fleurus, and made them masters, in a few days, of Brussels.

Reaction against the Reign of Terror.
By this time the horrible excesses of the Reign of Terror had begun to provoke the reaction certain at last to set in; and the triumphs of the Republic concurred in making the system of Jacobinism, at its worst, disliked. The conscience even of the populace of the towns revolted at the scenes of blood and despair which had made France miserable in the midst of her glories; and a growing sentiment quickly spread that the discomfiture of the enemy on the frontier ought to bring to an end a state of things which had brought such frightful confusion and havoc. The judges of the Revolutionary Tribunal sickened at their cruel and execrable work; shouting crowds no longer followed the guillotine; and cries of pity often rose for the victims even in the least wealthy parts of the capital. In this condition of opinion the ultimate fall of the supremacy of Robespierre was assured; but it was accelerated by a movement in the governing powers which had bowed under his sway for a time. In a fit, apparently of moody discontent, he absented himself for several weeks from the ruling Committee of Public Safety; and whether he did or did not contemplate the decimation of the down-trodden Convention, the execution of most of his nearest associates, and an absolute dictatorship for himself, most of his colleagues began to combine against

him. When he re-appeared in the Convention, the Fall of Robespierre, July 27, 1794. dark threats he uttered seemed to indicate only more measures of blood; and, under the influence of one or two courageous leaders, even the prostrate Assembly broke out in murmurs. Next day, after a scene of violent excitement, his arrest and that of St. Just and Couthon was decreed; and the Revolutionary Mountain at last rose with the Plain and Right against the dreaded tyrant. Robespierre, however, had in the interval invoked the aid of the Jacobin Club and of his satellites in the Commune of Paris; and he was rescued, with the two other prisoners, while a formidable insurrection was set on foot to overawe the national representation. The sections were, however, divided; a small part only obeyed the Commune; and the majority sided with the Convention, especially after a decree had been made declaring 'the triumvirs' traitors to the State. Robespierre and his associates were quickly Execution of Robespierre, St. Just, Couthon, and others, July 28, 1791. haled before the tribunal which, so to speak, had become the type of their fearful government; and most of the leaders of the Commune, now again struck down, perished with the abhorred and guilty tyrant. This apostle of blood and his followers were the last of the band, with a few exceptions, which was most stained in the Revolution with crime, and the dagger of Charlotte Corday[1] had

---

[1] Charlotte Corday, born in 1768, was a young lady of a good family in Normandy, and was a grand-daughter of Corneille. Her

some months before relieved France of the presence of Marat.

Reflections
on this
event.
Such was the Revolution of July 1794, or of Thermidor, by the new French calendar. It will always be a subject of reproach to Frenchmen that they bowed their necks to the yoke of Robespierre; and in this acquiescence we, no doubt, see the national tendency to yield to despotism. It must be recollected, however, that the success of the Jacobins was largely due, in the first instance, to its association with the cause of the independence of France, and to the hold they had on patriotic minds, and that it is impossible at a terrible crisis to check even the worst tyranny at once; and when the danger of foreign war had ceased, the Reign of Terror soon came to a close. As for the horrors of that time, they show how fierce were the hatreds of class which had long existed, and how brutalised a part of the people was; but though France accepted the Jacobin rule, and even welcomed it for some reasons, these atrocities ought, in justice, to be charged against a minority of Frenchmen only—the worst populace of a few great cities, and a band of reckless and auda-

The Terror-
ists were
not able
men.
dious demagogues. The Terrorists have been described as men of great powers, and the measures

imagination, deeply impressed by the atrocities of the Reign of Terror, fired her to assassinate Marat, and she stabbed him in a bath in July 1793. Her execution is touchingly described by Mr. Carlyle.

they adopted for the defence of France have been held up as a proof of ability; but this misconception of the worshippers of success ought to be contradicted by impartial history. The Jacobin leaders, certainly, showed energy; but their system led to a civil war which was destructive, and might have been fatal; their policy of force, especially in its social aspects, was cruel, ruinous, and unwise alike; and whatever seems to have been achieved by them was really achieved by French genius and valour. Besides, any credit to be given to them ought to be confined to Danton alone—the Marats, the Robespierres, and their crew, were simply incapable as political chiefs; and not one of the distinguished soldiers who appeared at this crisis was a Terrorist. The efforts of France to resist her foes were heroic, and have hardly, perhaps, been ever surpassed, but should not blind us by false illusions. The Allies might, without the least difficulty, have entered Paris in the summer of 1793; and, memorable as its struggles were, the Revolution triumphed only through the divisions and negligence of its antagonists. Nor does the eventful contest of this period detract from the truth that armies of recruits are weak and dangerous instruments of war, and that in the military, as in other arts, experience and training are of the greatest value. The young French levies were for months useless unless supported by seasoned troops. Napoleon, indeed, has said that what was really done

Notwithstanding the efforts of the French, the Allies could have put down the Revolution.

was done by the Army of the old Monarchy; and the forces of the Coalition were, in every respect, of better quality than their opponents. But mere organization is not everything in war; and unanimity, numbers, patriotic devotion, and above all, superior strategic skill—mistaken as Carnot was more than once—prevailed as they have prevailed before. These considerations ought not to lessen the admiration which is justly due to the energy and constancy of the French people; they simply explain, on rational grounds, the great success of the imperilled Republic, which national enthusiasm has not unnaturally invested with a character of marvel.

CHAPTER VII.

THERMIDOR. FRENCH CONQUESTS.

THE authors of the Revolution of Thermidor had Reaction of Thermidor. no conception that what they were about to do would bring the Reign of Terror to a close. Some had been almost as bad as their victims; others were Jacobins of a decided type; and their principal object was to escape death, though the majority of the Convention felt nobler motives. But the fate of Robespierre was a signal for France to throw off a terrible incubus; and a reaction against the Reign of Terror began to set in with that passionate quickness which is a distinctive feature of the national character. Within a few days the astonished multi- The prisons opened. tudes of 'suspects' were let out from their prisons; and even the populace of Paris joined in the ecstacy of the hour of deliverance. Before long the atrocities in the South and other places caused general indignation, and several of the monsters who had encouraged these crimes met the fate which they righteously deserved. After a time, too, the Punish- ment of several of the Terror- ists. Revolutionary Tribunal, with its detestable procedure, disappeared; and some of the judges justly

Abolition
of the
Revolu-
tionary Tri-
bunal,
perished by the violent means which they had recklessly abused. Meanwhile the Convention, at last
set free, endeavoured to confirm its restored supremacy, to check tyranny and anarchy alike, and to
inaugurate a policy of conciliation. The powers of
the Committee of Public Safety were reduced, and
its members changed by a speedy rotation, though
this obviously weakened the Executive. The decrees
which placed all France 'in requisition' to the State

and of the
maximum.
The forcing
of the value
of assignats
discon-
tinued.
were either modified or repealed; and the maximum
was abandoned, with the sanguinary laws which
sought to force the value of assignats, although the
results were not unforeseen. At the same time
energetic efforts were made to curb and guard
against mob license; the National Guards were
again remodelled and recomposed from the middle

The popu-
lace of Paris
kept down.
classes; the bands of pikemen were broken up; the
authority of the Commune of Paris, already shattered, was still further lessened by dividing its
council and limiting its powers; the more violent
sections were jealously watched; and, last and most
important of all, the revolutionary committees were

The Jaco-
bin Club
suppressed.
everywhere suppressed, and the Jacobin Club and
its kindred societies, the centre and feeders of agitation, were shut up. The remains, too, of the proscribed Gironde, with the seventy-three imperilled
deputies, were invited to return to their seats; compensation was voted, to a certain extent, for some of
the worst outrages of the Reign of Terror; and at

last Billaud Varennes, Collot d'Herbois, and Barère,
the three surviving chiefs of the terrible committee,
were prosecuted and sent beyond the seas, though
they had taken part with the men of Thermidor.
Finally, religion was solemnly declared free, and the
churches were given to their congregations, though
the sentiment of the Convention remained hostile
for the most part to priests of all kinds.

In this way the State tried to atone in some
measure for the horrors of the past, and the machinery of Jacobin disorder and cruelty was, to a
considerable extent, destroyed. The reaction, however, in the ruling powers of France, and the enact- Violence of
ments sanctioned by the Convention, expressed but the Reac-
feebly the intense hostility which broke out generally tion.
against the whole scheme of Terror. Jacobin functionaries were expelled from their places everywhere;
the National Guards of Paris, filled with the bour-
geoisie, showed no mercy to the 'tools of Robespierre;' and the young men of the Middle classes
formed companies to keep down the mob, and hunted
out, as if they had been unsexed, the female furies
of the galleries and the guillotine. The example of
the capital was followed elsewhere, especially in the
large trading cities, which had been treated with
such ruthless barbarism; and the recoil of opinion
was so quick and violent that the royalists, who, a
few months before, lived in daily dread of a sum-
mons to the scaffold, showed themselves, and some-

times oppressed their oppressors. The 'Committee of Mercy' into which, it was said, 'France had suddenly resolved herself,' was, in a word, not merciful to the late dominant party; and, in the rapid oscillation of the public sentiment, not only was clemency lavishly displayed, but the tyrants of the other day received their own measure, and were widely subjected to no little tyranny. At the same time, in Paris and elsewhere, a singular revolution in manners took place, not unknown in other national crises, but strangely rapid and very characteristic.

Revolution in manners.

In the extraordinary confusion of the last two or three years property had changed hands to an immense extent; and a new and large moneyed class had sprung up, formed by the sale of the lands of *émigrés*, by army and other government contracts, and, above all, by jobbing in assignats, and speculating in their continual fall, which no policy of terror could long prevent. This class, persecuted by the Jacobin leaders, now emerged brilliantly to the surface; and, the Court and the Nobles having disappeared, it formed the high social life of the capital, and stamped its character on the fashion of the hour. The uncouth savagery which had been supreme was replaced by a costly display of wealth; and the ruling orders banished the memory of the past in a giddy round of excitement and pleasure. The mansions of the Soubises and the Noailles were crowded with a new kind of *noblesse*, and echoed to

the sound of *bals à la victime,* confined to the relations of recent sufferers.   What was significantly
called the *jeunesse dorée* of the changed era appeared
in the salons of the Voltaires, the Condorcets, the Du
Deffands; and the wives and daughters of the men of
the time, in Ionic garb, and with snooded tresses,
aped the graces, the luxuries, and the dissoluteness
of Versailles.   The raggedness and austerity of 1793
was, in short, cried down; and French nature, vola-
tile and gay, indemnified itself for what it had
endured by rushing wildly into joyous amusement.
The change was not surprising, though it leaves behind a painful impression of national levity; yet we
shall hardly compare it, as it has been compared, to
the reawakening of nature in spring, to the letting
loose of the ice-bound waters.

It was impossible but that this vehement reaction should lead before long to renewed troubles.
The party of Terror, lately all powerful, had still a
considerable hold on the masses, though its chief
strength had departed from it; and the harshness
with which it was everywhere coerced, and the
triumph of the Moderates, now again in the ascendant, filled it with resentment and indignation.
Had France, 'patriot' orators exclaimed, shaken off
an arrogant though abhorred despotism to fall into
the hands of money-changers and scribes? Had
Europe been driven from her frontiers, and thousands of her bravest children perished, to substitute

for an aristocracy of birth and titles a new aristocracy of the bank and the counter? Was the
end of the Revolution to be the complete destruction of its most trusty instruments? Were the
measures by which it had saved the Nation, checked
dangerous factions, and maintained the poor, to be
flouted in the interest of the selfish and rich? Was
a dictatorship, stern, perhaps, but glorious, to be
converted into a mode of government in which a
class maltreated and scorned the people? The extraordinary condition of France gave plausibility
and force to these arguments, and supplied discontent with its keenest stimulants. The requisitions
and spoliations of the Reign of Terror had inevitably
lessened and checked production; and the abolition
of the maximum and of the ferocious laws which
forced up the value of assignats had concurred to
raise the price of commodities, though these expedients had, of course, been less efficacious than
their authors supposed. The result was that great
scarcity prevailed, and that a sudden and extreme
increase in the cost of the necessaries of life took
place; and the pressure in Paris became so alarming
that the government was obliged to put the poor on
rations, and to have recourse to all kinds of expedients to secure for them a scanty subsistence. This
distress, general and widespreading, caused a demand
for the Jacobin measures to revive; and it is probable, indeed, that the financial system of the Terror-

Scarcity
and dis-
tress.

ists, execrable as it was, was abrogated with incautious celerity. However this may have been, the lower classes in the capital and other parts of France lent themselves before long to the appeals of agitators to rise and regain their lost power; and the irritation they felt was, no doubt, exasperated by the selfish luxury of the new-made rich, by revolutionary hopes not yet extinguished, by ignorance, jealousy, and blind passion.

The Jacobin party tries to rally.

Such, briefly, was the internal state of France within a few months after the Revolution of Thermidor. The forces of anarchy before long broke out in the chief centre of their power, though they made themselves felt in other places, especially where they had been most repressed. On April 1, 1795, 12th Germinal by the new style, the mob of Paris burst into the hall of the Convention, shouting for 'Bread and the Constitution of 1793,' which had become the rallying cry of the 'patriots;' but it was driven out without much difficulty; and the dispersion of it was chiefly remarkable in that Pichegru, then for the moment on the spot, was called in to put down the rioters—an ominous but significant symptom. Some weeks afterwards, on May 20, or 1st Prairial, a more determined, and better organised demonstration took place; the populace, aided by one or two of the sections, invaded the seat of the Legislature again, and savagely massacred one of the deputies, amidst a scene

Outbreaks of 12th Germinal April 1,

and of 1st Prairial May, 1795,

worthy of the worst days of 1793 ; and a few Mountain deputies, who, it is supposed, were privy to the rising to some extent, went through the form of voting decrees which conceded all the anarchists' demands. This outbreak, however, threatening as it became, was no longer sustained by the potent means ready in the hands of a Danton or a Robespierre, and was suppressed in a short time ; and the National Guards and anti- Jacobin sections were again aided by a force of soldiery, now on the side of authority and the Sate, not as had been witnessed a few years before. The extinction of this insurrectionary effort enabled the leaders of the Convention to strike down the remaining Jacobin chiefs, and to take severe measures against future disorders. The deputies of the Mountain who had voted for the

put down and suppressed.

decrees were executed, or put an end to themselves ; and the relics of the Terrorists were prescribed and banished. At the same time the rebellious sections were disarmed ; the National Guard was carefully thinned of men suspected of the Jacobin taint, and, for the first time, was, to some extent, placed under regular military control ; and provision was made for the immediate removal of the Convention to Châlons in the event of danger, and for summoning to its aid the nearest army. Meanwhile, stern and sanguinary laws were passed against popular and anarchic meetings ; the 'patriots' complained that they suffered more than they had ever inflicted in the Reign of Terror; and, in the words of a sober historian, 'the party of humanity and moderation did not itself abstain from the profuse shedding of blood.'

By these means the once terrible power of Jacobinism was altogether broken, though its elements retained indestructible life. The government, however, had no sooner put down one party than it found it necessary to restrain another, for the reaction of Thermidor was becoming dangerous; and though the Moderates in the Convention prevailed, they had no sympathy with the avowed royalists, or even with the reformers of 1789, foremost in the fierce anti-Jacobin crusade. Coercive measures were also employed against these enemies of the Republic; and thus the ruling powers were on either side beset by exasperated and reckless factions, and with difficulty kept a middle course between them. The government accordingly became weakened; its authority, diffused and no longer concentrated, through the change made in the Supreme Committee, grew vacillating and, in a great degree, uncertain; and as it rejected the expedients of the Reign of Terror, it was gradually more and more compelled to look to military force for support—the end to which things were beginning to tend. Meanwhile, on all points in the theatre of war, the success of the French arms had multiplied, and the hosts of the Republic were borne forwards on a rapid and overwhelming tide of victory.

The power of Jacobinism finally broken.

Measures of the Government against the Royalists.

Weakness of the State. Tendency to the rule of the sword.

Great successes of the French against the Allies.

The fortresses captured in 1793 were quickly evacu-
ated by the Allies; and, after the occupation of
Brussels, the conquerors spread over Belgium in
triumph, and annexed its fertile provinces to France.
Before long Pichegru advanced northwards, while
Jourdan turned towards the Lower Rhine; and
though this dislocation of the French armies—a
characteristic error of Carnot's strategy, which consisted in ambitious movements on the wings of the
adversary with a too feeble centre, and was only
better than the impotent system of a general advance
on an immense divided front—gave the Allied commanders a great opportunity, they separated from
each other in eccentric retreat, full of mutual discontent and suspicion. By the close of 1794 Pichegru
had overran a large part of Holland, while Jourdan
had gained two important victories on the principal
affluents of the Lower Meuse; and within a few

Conquest of
Belgium
and Hol-
land, Sep-
tember,
1794,
January,
1795.

months the United Provinces had been transformed
into the Batavian Republic, the House of Orange
had been deposed, and the whole Low Countries,
from the Scheldt to the Ems, had become merely a
French dependency. The war, too, had been carried
far into Spain; and events, which for a time had worn
a menacing aspect in La Vendée, turned in favour of
the Republic once more. In this unhappy region
the atrocious cruelties of the Terrorists had caused
the insurrection to revive, and to extend over a large
part of Brittany; and the prospects of the rising

appeared so bright that an English expedition was despatched, with a band of *émigrés*, to aid the royalists. A descent, however, attempted from Quiberon Bay, proved a miserable and inglorious failure; and Hoche, who, like all real Generals, had many of the highest gifts of a statesman, reduced the whole West before long to submission by a policy of conciliation and sagacious firmness, winning the purest fame of the military chiefs of the time. Failure of English descent from Quiberon Bay, July 15-20, 1795.

These extraordinary successes of the French dissolved the already yielding Coalition. Prussia, the Power which had chiefly provoked the contest, was the first to abandon the allied cause, and made peace in the spring of 1795. Spain followed her example within a few months; and England, Austria, and Piedmont, with some States of the German Empire, already wearied of a calamitous and unprofitable struggle, alone remained to continue the war. The Republic had thus in two campaigns broken up an alliance which seemed more powerful than that which had humbled Louis XIV.; and it had extended its conquests beyond the limits of the most ambitious hopes of the Bourbon Monarchy. The result was in the highest degree brilliant; yet its real causes may be easily noted. Before the Campaign of 1794 had closed, the French armies, already immensely superior in numbers to their antagonists, had become gradually inured to war, and the young levies, after enormous losses, had hardened into truly The Coalition dissolved. Prussia and Spain make peace, April, June, 1795. Causes of this astonishing success of the Republic.

M

-formidable soldiers. The enterprise of the Republican troops, stirred by the impulse which first gave them strength, by the national passion for military glory, and by reiterated and splendid success, became astonishingly great and daring, and they ultimately gained that moral ascendancy over their ill-led and beaten opponents which is one of the chief conditions of success. The conduct, too, of the Allied Commanders was even more pitiable than before; and indignation was justly felt in England against the incompetent Duke of York, and in Austria against the dull Prince of Cobourg, who had contrived in two-years to fail in everything. The circumstance, however, has yet to be mentioned which so quickly enlarged the conquests of France, and we shall see it again in operation. The Republican soldiers were not, indeed, always kindly masters in the Low Countries or elsewhere; they were obliged to live on the tracts they occupied, being almost destitute of supplies from home; and their rapid advance was usually marked by excesses of license and by organised plunder. But in these, and in other parts of the Continent, the abuses of Feudalism and of the eighteenth century had undermined the whole frame of society; and the old order of things collapsed when it came in contact with revolutionary passions. Wherever the arms of France made their way, the privileges of the Church and the Nobles disappeared; the Reign of Liberty and Equality was proclaimed, and much that was unjust was swept away ; and the result was that the people welcomed the foreign invader in many places, though their liberation cost a heavy price, and that the moral influence of the new. French ideas was even more decisive than what were called the fourteen armies of the French Republic.

While France, however, was triumphant abroad, her government [1] at home remained feeble, and her social condition was in many respects lamentable. Her armies, indeed, with the exception of that which held the mountainous line of the Alps from Dauphiny and Provence to the Genoese seaboard, were, on the whole, in a prosperous state, especially in the rich Low Countries ; and the attraction to them became so great that towards the close of 1795 she had probably four hundred thousand men in the field. The peasantry, too, were for the most part thriving; notwithstanding the late maximum and requisitions, for the emancipation of the soil in 1789 had continued to make agriculture improve, and rents and taxes had sunk to almost nothing, under a currency

Continuing weakness of the Republic at home.

---

[1] M. Thiers' *Histoire de la Revolution Française* seems to me, on the whole, the best guide for the period between the Revolution of Thermidor and the 18th Brumaire. His account of the internal and financial state of France during these years of disenchantment and exhaustion is lucid and able. The papers by Napoleon, in his *Commentaries*, on Vendémiaire and La Politique du Directoire, should also be read, but they are not just to the Government. The correspondence of the late Mr. Wickham throws much light on the relations between Foreign Powers and the discontented factions in France.

ever diminishing in value. Trade, too, had revived
to some extent, the Reign of Terror having ceased
to destroy it; and the assignats, from their enormous
fall, becoming almost useless as instruments of exchange, a return to a natural system had begun, and
the precious metals slowly reappeared. But, as if to
show the irony of fate, the populace of the great cities,
which had figured so largely in the Revolution, re-

Extreme
distress of
the great
cities.

mained generally in extreme want; and though, as we
have seen, a moneyed class had sprung up, this had been
at the cost of other large classes; and the government,
which still went on receiving the imposts of the state
in worthless paper, was on the verge of financial ruin.
Harassed, too, as it was by contending parties, and
itself a mere revolutionary growth, its weakness could
only rapidly increase; and, with the Convention, it
was completely eclipsed by the splendour of the
military power, which had begun to fascinate the
masses. A strong Republican spirit, indeed, was
still prevalent in the legislature; but though freedom
and the Rights of Man were potent spells of victory
abroad, they were gradually losing their magic in

Exhaustion
of the revo-
lutionary
spirit.

France. The period of exhaustion and of disenchantment which follows revolutions was soon to open,
and the political aspirations of many turned chiefly
to repose and a strong government. What that
government would probably be in the collapse of
settled authority and rule, Burke in England had
already distinctly foreseen.

As the summer of 1795 progressed, the reactionary parties increased in strength. The Republic, though victorious abroad, became associated in the minds of thousands with Jacobinism and the horrors of the past; and a sentiment began to be widely diffused in favour of Monarchy and of the system which had perished only in a moment of passion. This feeling allied itself with the desire for quiet which largely prevailed; and though the *émigrés* were generally hated, and the exiled Bourbons had not many supporters, royalist agents made their presence felt, and the air grew thick with rumours of royalist plots. The government and the Convention, too, became more than ever disliked; they were accused of prolonging an usurped power; and as they had lost their hold on the Jacobin 'patriots,' they were decried in the centres of public opinion, though still upheld by the great mass of the Nation. In this state of things the ruling powers resolved not unwisely to appeal to the people; and the appeal was prefaced by a Constitution which expressed the latest effort of their legislative wisdom. This scheme, called the Constitution of the year III.—the Hegira of 'liberty' ran from 1792—plainly showed what were the ideas dominant among the chief French politicians of the hour and in the majority of the Convention. The organic changes of 1789 were ratified by a solemn oath; the Jacobin Constitution of 1793 was pronounced impossible and thrust aside;

Desire for repose and a settled government. Reaction towards Monarchy.

Constitution of the year III.

and the government was declared a Republic, though not without one or two protests. It was sought, however, to provide against the troubles and disasters of the past by a variety of ingenious expedients; and the proposed form of government was, in many respects, decidedly hostile to democratic influences. The Legislature, composed of seven hundred and fifty deputies, was to be elected by a not popular vote, although the election was to be annual; and it was divided into two distinct parts, a Council of Ancients, and one of Five-Hundred, experience having already taught the lesson of the perils attending a single Chamber. An Executive was formed of Five Directors chosen by the Councils, and with dependent ministers; and precautions were taken against a recurrence of the tyranny of 1793 by a provision that one Director should retire each year. At the same time, the extravagant local powers which had been created in 1789, and had been so terribly abused, were still further limited; and recent enactments against mob violence were declared essential to the security of the State. In addition, and most important of all, two-thirds of the existing Convention was to be re-elected, and a third part only of the succeeding assemblies was to be at present formed of new members, the mischief of the self-denying ordinance of 1791 being fully understood, and great apprehension being felt of the royalist and anti-republican parties.

This Constitution, which, in the abstract, was not without considerable merit, and might have struck root in different times, was generally well received in France, though it was observed that assent was for the most part passive, and the enthusiasm of past years had died away. The conditions, however, which maintained the existing Legislature in the main unchanged, were violently denounced in several places ; and this was eagerly seized as a grievance by the adversaries of the existing order of things. The leaders of the reactionary parties declaimed again the tyrannous Convention ; and they were supported by an undefined following of those whom vanity, ambition, and want, led to hope for advantage in new disorders. These sentiments were especially strong in Paris, ever agitated and eager for change ; and a formidable opposition to the Constitution, supported largely by the middle classes, gathered in the fickle and excitable capital. An insurrection was planned in the sections in which the malcontents were most powerful ; and on October 4 the National Guards of one of the principal sections rose, the expedients of anarchy being thus employed, in the turbulence of revolutionary time, by the class which had lately most suffered from them. The incapacity of the military commandant in Paris led quickly to a more general rising ; and on the morning of the 4th dense columns rolled through streets and squares towards the Tuileries palace, vociferating against

The Constitution generally well received.

Opposition to the re-election of two-thirds of the Convention.

Rising of the reactionary sections of Paris, 13th Vendémiaire, Oct. 4, 1795, put down by Bonaparte.

'Conventional traitors.' The insurrection appeared as terrible as that of August 10; but a man of action was on the spot to quell it, and the conditions of the struggle were wholly different. The frightened Convention had some hours previously given Bonaparte the command of all the troops in the city; and that officer awaited the attack with composure, though he had only then a few thousand men. The tumultuary assailants were cut down by vollies of grape shot as they appeared; their masses, after a few discharges, broke, and in a very short time hardly a trace remained of what had seemed a most alarming outbreak. The result was, perhaps, in some degree due to the energy and skill of Bonaparte; but probably the greater part of the force of the sections had no real heart in the cause; and as revolutionary passions were dying out, and regular soldiers were now on the scene, the revolt was put down with comparative ease.

The authority of the Convention restored.

The quick suppression of this outbreak, known as that of the 13th Vendémiaire, was a severe blow to the revolutionary parties, and, for the moment, put them to silence. The authority of the Convention and of the Republican chiefs, who guided the majority, increased in proportion; and severe measures were adopted against the still formidable National Guard of the capital. This citizen force was in part disbanded, and placed entirely in the hands of the General in command of the regular

troops in Paris; and it thus finally lost the character
of a power self-elected and independent of the State.
The chief result of Vendémiaire, however, was, of The military power
course, to strengthen the military power; and Bona- becomes
stronger.
parte learned on that day a lesson he was not likely
to forget. On October 26 the Convention declared
its mission ended, and closed its sittings; and
immediately afterwards the new powers which were
to govern France were installed in their functions.

The last part of the rule of the Convention is not less Reflections
on the
instructive than that which preceded, though of less course of
the Revolu-
tragic and striking interest. In less than a year tion after
Thermidor.
and a half after the Revolution of Thermidor, the
national sentiment seemed transformed; the forces of
Jacobinism had been put down; and the Republic
was threatened by a combination of royalists and
anti-republican parties, increasing in strength though
not dominant. In this we certainly see clear proof of
the mobility of the French character; yet, if we
recollect that the excesses of the Reign of Terror
were justly abhorred, that the ascendency of Jacobinism was largely due to passions engendered by
national peril, and ceased when the crisis passed
away, and that old habits, traditions, and beliefs
retain always extraordinary power, the change becomes intelligible to thoughtful minds. Concur-
rently we observe how the hopes and passions of the
Revolution begin to fade and wane in the disappoint-
ment of its most eager supporters; how the State

weakens amidst the strife of factions, not morally
strong, but selfish and fierce; how a feeling of
lassitude creeps over human nature lately so violently
stirred, and a desire grows up for rest and order;
and how, above all, the power of the sword, sustained
by brilliant success abroad, and throwing its weight
into the balance at home, casts its shadow on the
coming time.

CHAPTER VIII.

THE DIRECTORY. BONAPARTE.

DURING the period we are next to survey, the French Revolution, losing its strength at home, and having triumphed over its foreign enemies, turns definitively into the path of conquest abroad; until, suddenly arrested by unexpected reverses, it collapses under the military rule to which it had been for some time tending. In the internal condition of France in these years we see the causes increasing in force which had been lessening revolutionary passions, and introducing the arbitrament of the sword. Under a system of government, not, indeed, as worthless as it has been described by the flatterers of success, but composed of men not of marked eminence, divided against itself, and sinking in repute, the strife of the dregs of factions becomes more vexatious, and the desire for tranquillity grows more general; and at last, after a long exhibition of weakness, violence, and uncertain counsels, the State falls into the hands of a great soldier, and national peril hastens the issue. This consummation—the usual end of epochs of wild and destructive change—is furthered by the advance of prosperity, and by the eagerness of the

Character of this period.

new interests formed by the Revolution to consolidate
themselves; and it is precipitated by the national
tendency to bow to power and military fame, and,
above all, by the splendid achievements and gifts of
the extraordinary man to whom France not unnaturally looked as her champion. But though the decline
of the failing Republic is of the deepest interest to
the political thinker, History, at this juncture, turns
her chief attention to the march of the Revolution
abroad, and to its contest with the old Powers of
Europe. There we see how the ideas of 1789,
though not so decisively perhaps as before, concurred
to speed the progress of the arms of France; and
how their influence was not unfelt even in the hour
of defeat and disaster. There, too, we see how war
assumes more ample and magnificent proportions,
under the impulse of a new and eventful time, and
the inspiration of commanding genius; and we,
mark, as Bonaparte appears on the scene, how he
alike extends the conquests of France and modifies
her Revolutionary foreign policy.

State of the
Republic
after Ven-
démiaire.

For some time after Vendémiaire, the internal
state of the Republic presented but few incidents of
striking interest. The suppression of the revolt of
the sections had, we have seen, quieted the reactionary parties; and though their foes, in consequence,
grew more daring, the efforts of these were of little
avail, and a Jacobin conspiracy, headed by an en-
thusiast called Babouf, came to nothing. Another

attempt was also made to descend on the coasts of
La Vendée ; but the policy of Hoche had borne its
fruits, and the West remained in submissive repose,
though elements of trouble lurked beneath the surface. The majority too, of the former Convention
predominated in the new Legislature, though a certain number of the freshly elected deputies were
little inclined to republican views; and, for the present, the prevailing sentiment was to maintain the
settlement of 1795. The Directory, who composed
the government, though less moderate than many in
the Councils, and indeed wholly formed of men who
had voted for the death of Louis XVI., agreed,
nevertheless, with each other for a time, and with
the national representation; and though, with the
single exception of Carnot, they were not men of
peculiar mark, their policy was rather mastered by
events than of a decidedly bad character. They
had, indeed, recourse to one or two expedients of a
Jacobin kind, in the exhausted financial condition of
the State ; and they levied a temporary forced tax on
the rich, and were compelled for a few months to
return to the arbitrary system of requisitions for the
troops. These measures, however, were soon abandoned, and were, perhaps, inevitable in existing circumstances ; nor can the Directory be fairly charged
with what ought to be ascribed to the tyranny of the
past. The same kind of excuse may be urged for
another startling and grave act, though not so iniqui-

Policy of the Direc-tory.

tous as it appeared. After many attempts to avoid
the catastrophe, the Directory, with the assent of
the Councils, deprived the assignats of their nominal
value, and declared that in all public and private
transactions they should be estimated only at their
real worth ; and before long, therefore, this degraded
currency disappeared wholly from circulation. This
was National Bankruptcy in another name ; but as the
Republic had not the means of redeeming the thousands of millions of notes afloat, and the State could
not exist on worthless paper, no other course was,
perhaps, possible; and, in any case, the men of this
era were not responsible for the original evil. Nor did
the abandonment of one of the last devices of a revolutionary age create general discontent; nor was
the shock as severe and ruinous as ˉMr. Pitt and
others supposed it would prove.   The assignats,
we have seen, had for some time been ceasing to be
a medium of exchange ; their depreciation had been
taken into account in all the ordinary dealings of
commerce ; and the ultimate loss by them was comparatively small, as their real value had sunk to
almost nothing.   What the paper system had done
was to transfer property to an enormous extent by
its diminution of fixed debts and payments, and
by the scope it gave to jobbing speculations. But
these mischiefs were now of old date; and the consequences, cruel and unjust as they were, affected
classes rather than national interests.

National
Bank-
ruptcy vir-
tually de-
clared.

Preparations were made to renew the war with increased energy in 1796. The military operations of the French, in the last months of the preceding year, had been unsuccessful to a considerable extent, for Jourdan had been driven from Mayence, and Pichegru had dealt treasonably with the enemy in his front ; and though a victory had been won by the French at Loano, upon the Italian seaboard, the tide of fortune ran less favourably for the Republic than it had ran before. Two large armies under Jourdan and Moreau were massed apart from each other on the Middle Rhine for a formidable invasion of Western Germany ; and a third, composed of about forty thousand men, good soldiers, but in extreme want, was entrusted to the youthful Bonaparte, and confronted—along the coast from Genoa to Nice, which it had occupied for a considerable time—a much greater Austrian and Piedmontese force. The vicinity of the Rhine, therefore, was to be the principal scene of events ; but the force of genius transformed the situation. Bonaparte assumed his command in the first days of April, and the presence of a superior mind was at once seen in the operations of the French. Deceiving his adversaries by rapid demonstrations, he quickly broke through their extended centre ; and, in a series of brilliant engagements, he divided the Piedmontese from the Austrians, drove both, routed, in separate retreat, and having, as he said, ' turned the Genoese

Preparations for the campaign of 1796.

The campaign of Italy.

Bonaparte invades Piedmont from the seaboard.

Alps,' reached Turin in a few days in triumph. He

April 28, 1796.
now made an armistice with the King of Sardinia, which placed the fortresses of Piedmont in his hands, and secured his communications with France; and having declined to revolutionise a State which might become favourable to French policy, he at once directed his whole efforts against the Austrians, who, he clearly perceived, were the only foes of real im-

May 9–30, 1796.
portance in Italy. Advancing with a celerity before unknown, he anticipated his antagonist, Beaulieu, on the Po; and after a murderous struggle at Lodi, he entered Milan, and overran Lombardy, the Austrian commander being unable to contend against such activity and daring, and being outnumbered in every encounter. At Milan Bonaparte was welcomed with delight, the citizens detesting the Austrian yoke, and being inclined to the new principles; but he halted only to strengthen his position; and having terrified into submission the hostile princes of Parma and Modena, he made straight for the

He marches to the Adige, Siege of Mantua.
line of the Adige, which he had marked out, with true military insight, as the real theatre on which to contend with Austria for the prize of Italy. Having forced Beaulieu across the Mincio, and compelled him to fall back on the Tyrol, he laid siege to Mantua in the first days of June; having previously refused, with equal prudence and firmness, to obey an order of Carnot to march against Rome,

which certainly would have led to disaster, by need-
lessly dislocating the French army.

Great as this success of Bonaparte was, the
cabinet of Vienna was not disconcerted, and made
vigorous efforts to repair its defeats. The French
army in Italy was known to be weak in numbers;
the strength of its position in the hands of a great
commander was not understood; and it was widely
believed that it was doomed to disaster, thrown
forward dangerously, as it seemed, round Mantua.
The beaten divisions of Beaulieu received large reinforcements from the Austrians on the Rhine; and
Wurmser, a veteran of high reputation, advanced, in
the last days of July, with an army that seemed
more than powerful enough to liberate Mantua, and
overwhelm his antagonist. Bonaparte, however,
with that rapid decision which is one of the distinctive marks of a great leader in war, forestalled
admirably the Austrian movements; and, raising
the siege of Mantua at a moment's notice, encountered his enemies as they descended along either
shore of the Lago di Garda; and, interposing between their divided masses, defeated them at Lonato
and Castiglione. He then turned to pursue his
baffled assailants, advanced boldly to the verge of
the Tyrol, and routed Wurmser again in the defiles
of the Brenta, after a march of extraordinary daring
and quickness; and though the tenacious Austrian
chief got into Mantua by a circuitous movement,

*The Aus-
trians send
an army to
raise the
siege of
Mantua,
and to
crush
Bonaparte.*

*He defeats
Wurmser
in a series
of engage-
ments.*

August 3-6.

*September
1-13, 1796.*

N

he had lost the greater part of a gallant army. The Austrian government, however, still persisted, and a

The Austrians send Alvinzi with a new army.

fresh force, under the command of Alvinzi, was once more directed against the adversary, who, it was thought, must succumb to such repeated efforts. The attack proved dangerous in the extreme; the Austrians, though disseminated in separate masses, forced one of the chief positions of their foes; and Bonaparte, with the main body of the French, was almost driven from the barrier of the Adige. Alvinzi,

He is defeated at Arcola, November 14–17, 1796.

however, paused at the decisive moment; his dexterous adversary fell on his rear, displaying the greatest fertility of resource; and victory at last declared for the French, after a protracted struggle along the dykes of Arcola. The campaign, nevertheless, was not yet ended; and after recruiting his worn-out host, Alvinzi again approached the Adige.

Decisive victory of Bonaparte at Rivoli January 14, 1797.

The decisive encounter took place on January 14, 1797; and the Austrians, divided and baffled once more, were routed with terrible effect at Rivoli, on the eastern shore of the Lago di Garda. This brought hostilities to an end for a time; Mantua opened its gates in a few days; and Bonaparte stood in triumph on the line which he had made the centre of his operations, having annihilated three armies, each stronger than his own.

Reflections on the campaign. Great skill of Bonaparte.

This splendid campaign, still perhaps unrivalled, raised Bonaparte at once to the summit of fame. Its astonishing results were in some degree due to

the revolutionary influence of France, but far more
to the capacity for war of the young leader who had
appeared on the scene. Bonaparte, in this admirable passage of arms, had displayed all the qualities
of a great captain: sagacity, resolution, boldness,
vigour, a perfect knowledge of the theatre of operations, and a skill in arranging his forces on it, which
completely bewildered his inferior opponents. There
was, too, another general cause for his success which
gave a character to his strategy, and has wrought a
marked change in the art of war. In consequence
of the multiplication of roads it had become possible to make more rapid marches than Generals of
a former age could attempt; and, owing to the progress which had taken place in husbandry, an army
could now often rely for supplies on the districts
which it happened to go through. The old system
of slow advances, depending mainly on magazines,
and retarded by fortresses and such obstacles, had
thus become, in a great degree, obsolete ; quick and
daring attacks and brilliant manœuvres, the troops
living on resources found on the spot, had been made
more practicable than they had ever been; and
Bonaparte had thoroughly grasped this truth, though
it had been partially recognised before, and it was
obviously suggested by examples already set by the
revolutionary armies. In this campaign we see
plainly that he conducted war upon these new principles ; and, though other causes no doubt aided, the

Character of his strategy.

N 2

circumstance partly explains his success in so often routing his adversaries in detail.   Nor had he shown the qualities of a soldier only in this memorable and arduous contest; he' had given proof of no common statecraft, and especially of a secret contempt for the propaganda of revolutionary ideas, of which the French armies were the principal centres.[1]   He had refused, we have seen, for political reasons, to overthrow the Sardinian throne ; and, to the astonishment of his lieutenants, he had soon afterwards negociated with the Pope and the Grand Duke of Tuscany, avowed enemies of the Revolution, in an anti-revolutionary and diplomatic fashion.   It had become already evident that a leader had appeared who, for good or evil, had little sympathy with the fanaticism of liberty and the Rights of Man, powerful levers, as yet, of French influence abroad.

State craft of Bonaparte.

Meanwhile a very different contest had been waging beyond the Rhenish frontier.   Following the essentially vicious plans of Carnot, Jourdan and Moreau had made their way into Germany, divided by a wide space of country, the first moving along the Thuringian range, the second skirting the Black Forest.   The young Archduke Charles retreated before them, though with an army nearly equal in

Campaign of 1796 in Germany. Defeats of the French.

The Archduke Charles.

---

[1] Napoleon's policy in 1796–7 is set forth by himself in his *Commentaries*. M. Lanfrey, in his *Histoire de Napoleon I.*, describes it with great ability, but in too harsh colours.   See chapters 6, 7, and 8 of the work.

strength; and he fought an indecisive battle with Ability displayed by him. Moreau at Neresheim, near the Upper Danube. As the French generals, however, moved slowly, and made no signs of effecting their junction, though now only a few marches apart, the Archduke assumed the offensive; and leaving a detachment to hold Moreau in check, he marched against that leader's isolated colleague, thus imitating the manœuvres of which Bonaparte was giving such splendid examples in Italy. The operations, however, of the Austrian commander were wanting in the perfect skill and energy conspicuously displayed by his far greater rival. Jourdan, indeed, was beaten in detail, and fell back discomfited to the Rhine; but he was not pursued with daring and vigour, and he reached his winter quarters comparatively unhurt. Moreau, too, after the defeat of Jourdan, got safely out of a most difficult position, and made good his retreat through the intricate defiles and rocky crags of the Swabian Alps; and he even drove his antagonist back, though probably had he been boldly attacked by the Archduke after he had been left without support, his army would have been almost destroyed. Germany was thus cleared of its French invaders, and the Republic met a decided reverse; but nothing really great was accomplished; and the campaign is a striking example how military conceptions, however excellent, must be as well executed to have marked results. The Archduke, however, justly acquired renown; the errors of Carnot were perceived, and the events of the year proved that in war, as in other arts, the same original thoughts, under similar conditions, occur sometimes to different minds.

The state of the Republic had not improved while Bonaparte had been conquering on the Adige. The prosperity of France, indeed, had gradually augmented, as time weakened the effects of the Reign of Terror; the distress of the great cities lessened; and the revenue had begun to show signs of progress. But, in the transition from the paper system, the finances were inevitably strained to the utmost; the part of the Debt remained unpaid which had escaped the Jacobin sponge; and the treasury was extremely ill-managed, as even the administration of it had been withdrawn by the Constitution from the Executive government. Complaints, therefore, abounded everywhere; and the animosities had only increased in bitterness, which threatened the State, and made it insecure. As Vendémiaire receded into the past, the royalists and anti-republicans grew in strength; and they drew to their party a great many of the disappointed and discontented men who always abound in a revolutionary time, and a still increasing number of the new aristocracy of wealth, who had no genuine republican tastes, and whose real aspiration was for rest and enjoyment. Dissensions, too, broke out within the Directory itself; and two of the Five, of

Internal state of the Republic; revival of factions.

whom Carnot was one, inclined at least to sentiments
opposed to those which prevailed in the old Convention ; ever true, in the main, to some ideal of a Republic, whether moderate or not. The strife of
factions was quickened by the elections held in 1797,
which displaced many of the ' Conventionals,' as they
were called—the restriction was by this time removed which had been imposed in 1795—and filled
the Legislature to a large extent with deputies of reactionary views.   The royalists and anti-republicans
of all kinds began now to assert their power ; and
the opposition to the government was seconded by
reckless and widespread intrigues and conspiracies.
Pichegru, whose royalist leanings had been avowed,
though his treason had not been yet divulged, be- Royalist
came a chief director of these dishonourable plots, and reac-
tionary
the inevitable growth of a revolutionary era ; and schemes.
though the reactionaries were, in the mass, opposed
to violent changes in the State, they lent themselves
to designing leaders.   Non-juring priests and long-
exiled *émigrés* began soon to return freely ; the
Powers at war with France had numerous agents in
correspondence with the malcontents ; and plans were
set on foot for a Bourbon restoration, to be sustained
by a rising in La Vendée.   In this emergency the
three Directors who adhered to the existing order of
things, sought for aid from the power alone capable
of throwing a decisive weight into the scale, and not
disloyal to their authority if not particularly attached

to it. Hoche moved an armed force to the capital, and Augereau, despatched from his camp by Bonaparte, was placed at the head of seventy thousand men to carry out the intended design. On September 4 the Tuileries palace, where the Councils held their ordinary meetings, was once more surrounded by bands of soldiers; and within a few hours most of the reactionary deputies were imprisoned and their seats declared vacant; and the principal conspirators, with Pichegru at their head, were on their way to a place of foreign banishment., The remains of the thinned and purged Legislature voted readily the proscription of many suspected persons, and one of the hostile Directors was included in the list, Carnot having fortunately effected his escape. The triumph

Coup d'état of the 18th Fructidor, September 4, 1797.

of the Republicans, in what was named the *coup d'état* of the 18th Fructidor, was for the moment general and complete; but the success of this and similar acts of violence—for which happily no word exists in our language—could only hasten the military domination which was already beginning to be felt everywhere. It was also a significant mark of the time that the populace of Paris, growing weary of political changes which had proved abortive, and of the struggles of warring factions, had looked on with passive indifference at the peril of the Republic and its temporary success.

Conclusion of the campaign of Italy.

Meanwhile Bonaparte had been extending the power of France in the Italian Peninsula, and, after

a brief and brilliant campaign, had brought the war
with Austria to a close. The first success of the
French on the Po had agitated the States between
the Apennines and the Alps; and national and
revolutionary passions had made the invaders welcome in many places, though sentiments were a good
deal divided, and the excesses of the so-called libera·
tors—especially the robbery of works of art, which
were sent as spoils to the museums of Paris—had
caused more than one angry rising. After the complete triumph of 1796 the opinion of the masses
became more evident ; and though the old aristocracy
of Venice remained bitterly hostile to the French
ideas, the Modenese, the subjects of the Pope, and
the great body of the people of Lombardy, had risen
against foreign or hated rulers, and attached themselves to the victorious Republic. Bonaparte, wield- Conciliatory policy
ing already enormous power, ably turned the move- of Bonaparte ;
ment to his own advantage, and to that of the his anti-
revolution-
Directory in a secondary degree; he obtained con- ary views.
siderable territories from the Pope, as the price of
sparing Rome and the adjoining Provinces; and
while he levied ample contributions from them, he
gave or promised the Italians ' liberty ' within the
districts he had annexed or occupied. He steadily
carried out, however, this policy of compromise, and
of moderating Revolution; and, while he treated
the Italian States and their Sovereigns with a view
rather to his own objects, or to immediate political

interests, than with the least regard to Republican notions, it was observed that he had no sympathy with what he contemptuously called the multitude, and that he thoroughly despised its hopes and pas-

He marches on Vienna from Italy. sions. Meanwhile, his army, largely recruited from all parts of France, had grown truly formidable; and he took the field again in the spring of 1797. Austria had no forces sufficient to oppose his march; and though the Archduke Charles made a gallant resistance, Bonaparte swept over the Italian Alps and hastened down their German slopes towards Vienna. An armistice was signed on April 7, within sight of the domes of the Austrian capital; and Bonaparte, having with a force comparatively small conquered from the Var almost to the Danube, and broken the strength of the Austrian Monarchy, dictated in a few months the terms of peace. By this treaty,

Treaty of Campo Formio, October 17, 1797. known as that of Campo Formio, Austria ceded Belgium to the French Republic, and, as head of the Empire, agreed to the cession of the German Provinces on the French bank of the Rhine; and she consented that Lombardy and several adjoining States should be formed into a Cisalpine Republic, of course a mere dependency of its French original. In return for these immense losses, Bonaparte flung her Venice as a spoil, notwithstanding a protest from the Directory; and his conduct in this was very

Gains of France. characteristic. The Venetian oligarchy had certainly been a thorn in his side while he was on the Adige; and after he had disappeared beyond the German

Alps it had stirred up an insurrection in his rear.
But, long before the peace of Campo Formio was
made, the Republic had become a democracy apparently subservient to French authority; and,
nevertheless, Bonaparte deliberately sacrificed a   Sacrifice of
people and a State, once an ally of France, in order,   Venice.
as he avowed, to sow dissensions among the late
Coalition, which, with the exception of the Power
aggrandised, resented the transfer of Venice to Austria.   The act was not so ineffably base as it has
been described by historical censors; but it was very
significant of a policy of craft, of expediency, and of
hard self-interest, opposed to all the revolutionary
professions.

In this manner a youth of twenty-seven had   Reflections
struck down the only remaining enemy feared by the   on the conduct of
Republic on the Continent, had consolidated and   Bonaparte.
widely increased its conquests, and had shed a glory
on the arms of France more splendid than she had
ever known.   The right of France to what the
national sentiment had recognised as her natural
limits had been admitted by her great German rival;
her influence extended, beyond, from the Adige to
the Texel; and a dream which Richelieu would have
dismissed as idle had been realised in perfect com-
pleteness.   A burst of enthusiasm went up from the   Enthu-
popular heart to hail the warrior who had done these   siasm in
France
great deeds; and the name of Bonaparte, scarcely   at his successes.
known before, was in every mouth in France as a
word of marvel.   Hardly less astonishment was felt

in Europe, too, at the extraordinary achievements of
the young conqueror; and the feeling was largely

Admiration felt for him in Europe. mingled with genuine admiration. The diplomatists
of Piedmont, Austria, and Rome, had recognised in
Bonaparte a kind of sympathy with the established
Powers and old order of Europe, surprising in a
negotiator of a Revolutionary State; and several of
them had said that no other General of the devouring Republic would have been so moderate. Bonaparte had also treated his defeated opponents with
delicate and becoming courtesy; and he had displayed to soldiers and statesmen whom he wished to
please the charm of a manner which possessed an
inscrutable and mysterious fascination. He was thus
an object of the respect and flattery of even the
most resolute enemies of France; and he was regarded by the enfranchised Italians as a deliverer all
the more to be loved because one of their own race
and blood.

He returns to France, and is received with acclamation. Surrounded thus by a halo of glory, Bonaparte
left Italy to return to France, and after passing
hastily through Rastadt, where the States of the
Empire were negotiating a peace that seemed inevitable after Campo Formio, he quietly returned to the
modest house in Paris which he had quitted a comparatively unknown soldier. He was greeted with
an enthusiasm such as never had been seen since the
days of Louis XIV., though, either from inclination,
or a studied policy, he avoided the public gaze, and

seemed to court solitude.   The capital shone in an array of splendour which contrasted strangely with the horrors of a few years before ; and the conqueror of Arcola and Rivoli was the only object in the eyes of the multitudes who crowded to celebrate his great exploits in festivals in which the antiqué pomp of the Roman Commonwealth curiously blended with the glitter and luxury of a modern age.   How long would the obscure Heads of a divided, feeble, and revolutionary government withstand the influence of the young hero, who seemed to carry fortune at the point of his sword ; how long could the Republic co-exist with this glorious personification of the military power, which already encompassed it on every side ?

CHAPTER IX.

EGYPT AND THE 18TH BRUMAIRE.

The Directory jealous of Bonaparte.

THE homage rendered to Bonaparte, and the great influence he already enjoyed, gave umbrage to the Republican government. The causes of dissension were already numerous, for the haughty independence of the young General, his contempt of all military schemes but his own, his sacrifice of Venice, and the sovereign attitude he had assumed in the negotiations with foreign Powers, and, above all, his supremacy over his troops, had been viewed with alarm and suspicion; and when, after his return to France, he was welcomed as the image of her glories, his ascendency irritated the eclipsed Directory. Nor did the subsequent conduct of Bonaparte tend to reassure the weak chiefs of the State, who dreaded an authority they did not themselves possess. Though he continued to live in extreme simplicity, and seemed to prefer the society of men of letters and science to political affairs, he had let fall expressions which revealed a dislike of a feeble and disunited government, and the junta in office instinctively felt that his presence was a rebuke and menace to them, though jealousy was masked under a show of deference. Either from a desire to get rid of a foe, or possibly from a higher motive, the Directory soon tried to engage Bonaparte in an enterprise which, if of tempting promise, was one of extraordinary difficulty and peril. England, after Campo Formio, was the only great Power that remained at war with the victorious Republic; and the Directory, exasperated at a recent failure to negotiate with a British envoy, invited Bonaparte to make a descent on our coasts, a project for which Hoche—that remarkable man had just died, amidst general regret—had always had a strong predilection. An expedition of this kind, however, had been unsuccessful in 1796, and the battles of Camperdown and of St. Vincent had annihilated the fleets of the Batavian Republic and of Spain, now an ally of France; and Bonaparte declared the scheme premature, and suggested another which he thought more hopeful. His mind, imaginative and calculating alike to a degree of force which has been seldom witnessed, had even in Italy turned to the East and the ancient centres of historic power; and he proposed to invade and occupy Egypt—that stage on the way from Europe to Asia which has always attracted the thoughts of ambition. The Directory joyfully sanctioned a plan which would certainly remove a dreaded rival, and, if successful, would make France predominant in the Mediterranean Sea; and Bonaparte was given ample

They engage him to attempt a descent on England.

He proposes to invade Egypt.

means to carry out the intended design. His preparations were made with a secrecy and skill which
showed a high faculty for organisation; convoys
were collected in the Italian ports, and troops
directed upon the sea coast, so as to conceal the
project as long as possible, and in May, 1798, the
expedition set sail from Toulon.   It consisted of a

Expedition
to Egypt.

powerful fleet and army; and its leader perhaps
entertained hopes of imitating the career of Alexander, and, after subduing and colonising Egypt, of
marching from the Nile to the Indus.

Congress of
Rastadt.

While this enterprise was being set on foot, the
Congress of Rastadt had been sitting, and negotiations were going on for peace on the Continent.
Prussia, which since the treaty of 1795 had almost
become an ally of France, had secretly rejoiced at
the defeats of Austria, and saw in the present confusion of Europe the means of extending her power
in Germany, fell in with the policy of the Republic;
and, in consideration of benefits to herself, assented
to French annexations on the Rhine, to the humiliation of lesser German States, and to the annihilation
of Imperial Bishoprics, a favourite object of the
Directory. This selfish and unpatriotic state-craft—
a main cause of that habit of aggression and of intervention in the affairs of Germany which Prussian
writers have of late laid to the charge of France for
their own ends—was opposed by many of the German
princes; but, as Austria had retired from the con-

test, and the divided Empire was left without a head, a renewal of the war appeared impossible. This would have been the case in ordinary times; but ancient privilege and democratic ideas were in a state of angry collision in the countries approached by the French revolution; and though hostilities had not broken out, the prospects of peace did not brighten. Causes of fresh troubles quickly arose when Europe was in this disturbed condition, and they were aggravated by the republican ardour and arrogant pretensions of the Directory, though the real impulse lay much deeper. Before Bonaparte had left Italy, Genoa had formed herself into a Ligurian Republic; and not long afterwards a democratic rising occurred in several of the Swiss cantons, and after a sanguinary civil war an Helvetian Republic had been established by French influence and French bayonets. This was followed by a violent outbreak in Holland, which for a time completely overwhelmed the party of the House of Orange, and of the old order of things; and before long the French invaded the territories of the Pope, set up a Roman Republic in his States, and filled Piedmont with revolutionary agents, against the conditions of recent treaties. The march of the Revolution hastened, accordingly, in a state of nominal peace as well as in war; and the French government encouraged its progress by their fanatical zeal and reckless want of scruple. It is not surprising, therefore,

Renewal of causes of discord in Europe.

Formation of the Ligurian, Helvetian, and Roman Republics.

when France had lost for a time her most dreaded commander, that several of the European Powers should have begun to watch events, and prepare for war; that the negotiations should have proceeded slowly; and that even Austria should have thought of arming once more, more especially as the genius and the gold of Mr. Pitt had been engaged in endeavouring to cement again the Coalition which had been recently dissolved.

Bonaparte lands in Egypt, July 1, 1798.The Continent was in this unquiet state when an unexpected event decided the issue to which affairs had been slowly tending. Bonaparte had reached in safety the shores of Egypt, the French fleet, though with immense convoys, having eluded the watch of the English cruisers, and having even had time to seize and occupy the great Mediterranean fortress of Malta. His army had landed, and, crossing the verge of the Desert, had routed the Mameluke horsemen in a battle fought within sight of the Pyramids; and he had triumphantly made his way to Cairo, where he had endeavoured to establish a French

Battle of the Nile, August 1, 1798, and destruction of the French fleet.colony. But in the meantime his fleet had been completely destroyed by the great English sailor whose manœuvres at sea bore a certain resemblance to his own on land; and he seemed cut off with his army from France, and imprisoned within his precarious conquest. The victory of Nelson determined the Powers which hitherto had been afraid to strike, and new names were added to the list of the enemies of the hated Republic. Hostilities were proclaimed in the winter of 1798, and it soon became evident that the contest would rage from the Zuyder Zee to the Straits of Messina, and would spread over part of the Turkish Empire. The Porte undertook to attack Bonaparte; the Court of Naples set an army on foot to invade the newly-created Roman Republic; Austria prepared for a fresh struggle on the Rhine and the Adige, aided by a large reinforcement from Russia, which had only threatened in 1793. Except Prussia, Germany generally concurred; and England gladly threw her sword into the balance. The Directory, elated by late successes, met the challenge of its foes with defiance, and looked forward confidently to a new series of triumphs. An unhappy incident which had lately occurred, and which threw a dark stain on the House of Austria—the murder of the plenipotentiaries of France at Rastadt—had given the Heads of the Republic the strength arising from widespread national indignation; and, as they had, so to speak, organised the *levée en masse* of a few years before, by the celebrated measure called the 'Conscription,' which at this moment is the foundation of the enormous armies that cover Europe, they prepared for hostilities on the greatest scale.

The campaign which followed is of little interest as an illustration of the art of war. On both sides the antiquated system of timid operations along an immense front, and of pausing at obstacles, was, in

Renewal of the war in Europe.

Murder of the French plenipotentiaries at Rastadt, April 28, 1799.

The Conscription.

Character of the campaign of 1799.

the main, adopted ; and Switzerland became the
chief scene of the strife, in deference to the wholly
unsound theory that the possession of a mountain
range ensures a decisive advantage to a belligerent,
apart from any other consideration. Though generally ill-led, the allied armies had for months a
great superiority over the French; and they certainly might have invaded France, and not improbably have occupied Paris, had they been directed
with real energy and skill. In the South, indeed,
the Neapolitan levies were routed with ease upon
the Tiber ; and under the impulse of the success of
their foes, Naples was changed into the Parthenopæan Republic, and the King of Sardinia was expelled from Piedmont. But on the points where
the contest was most important, fortune was long
adverse to the French armies ; and they lost the
fruits of the glorious struggle of 1796, though
ultimately saved from the extreme of peril. Jourdan
was defeated with heavy loss at Stochach, between
the Swabian Alps and the Lake of Constance ; and
had not the Archduke Charles been compelled, by
the military or Aulic Council at Vienna, to waste
his strength among the hills of Uri, he might have
crossed the Rhine, invaded Alsace, and turned the
whole line of the French in Switzerland. Meanwhile the Austrians, feebly resisted, had forced the
great barrier of the Adige, and before long the
warriors of Mantua and Rivoli were driven from

Defeats of the French.

Formation of the Par-thenopæan Republic.

Battle of Stochach, March 25, 1799.

the Mincio across the Adda, pursued by the enemies they had so often beaten, and by a Russian army under Suwarrow, a celebrated veteran of the reign of Catherine. Moreau, who had been for some time in disgrace, for supposed complicity with the crime of Pichegru, which had come to light after the 18th Fructidor, was now raised to command in Italy, and endeavoured to effect his junction with Macdonald, coming up from the South across the Apennines ; but though Suwarrow showed no skill, the two French generals were completely beaten along the historic banks of the Trebbia. This was followed by another defeat at Novi ; and though the populations of the new Republics remained true to the French cause, the allied armies overran the Peninsula, and Italy was lost more quickly than it had been won, with the exception of Genoa and a few other fortresses. The war might have been easily carried into France had the allies now acted with real vigour ; but a fatal error caused a sudden change of fortune. A combined English and Russian force, under the too celebrated Duke of York, had made a descent on the coasts of Holland; and the Archduke Charles was directed, on the Lower Rhine, to co-operate with this remote detachment. This movement, made against the will of the Austrian chief, weakened the force opposed to the French in Switzerland; and Massena, the ablest lieutenant of Bonaparte, and trained in the lessons of 1796, seized

Battles of the Trebbia and Novi, June 17, 18, and 19, and August 15, 1799.

The French driven from Italy.

Failure of English descent on Holland.

Battle of
Zurich,
September
25–28,
1799.
the favourable opportunity presented to him. He
fell on the Russian Korsakoff in his front, and
crushed him in a great battle at Zurich; and
Suwarrow lost three-fourths of his army in a fruitless attempt to support his colleague. This reverse
as usual, caused dissensions to break out in the
camp of the allies, and these were only increased by
the inglorious failure of the Duke of York in his
advance into Holland, which the Archduke's diversion
It saves
France
from inva-
sion.
could not really aid. Offensive operations were
given up, and the territory of France remained
intact; though the armies of the coalition, with
Italy in their grasp, had their outposts on the
borders of Provence.

Lament-
able inter-
nal state of
the Re-
public.
Meanwhile, torn by intestine factions, the government of France had been declining rapidly; and the
state of the Republic had become lamentable. The
coup d'état of the 18th Fructidor had given a
triumph to the extreme republicans; and the expir-
ing remains of the Jacobins lifted their heads again
Strife of
factions.
in a threatening manner. The Directory and the
Councils, becoming alarmed, turned violently against
the enemies they feared; and several 'patriots' of a
Jacobin type having been returned at the election of
1798, the reckless course was again taken of declaring the seats of these deputies vacant, as had been
done in the case of the opposite party. The Constitution was thus set at naught twice; and though
the conduct of the ruling powers was less to blame

than at first sight appears, the Republic became
more feeble than ever, and degenerated into a
divided oligarchy, discredited, unpopular, and merely
upheld by the military force. on which it rested.
The renewal of the war in 1798 gave extreme
offence to the wealthy classes, and roused once more
anti-republican hopes, though the fate of the envoys
at Rastadt had, we have seen, provoked a storm of
indignation; and measures on which the Directory
unwisely ventured—a renewal of the forced tax on
the rich, and a declaration which practically swept
away the greater part of the remaining Debt—caused
widespread irritation and alarm. At this crisis the
reverses of 1799 came to exasperate passion and discontent, led to fresh exhibitions of weak oppression,
and precipitated the decline of the imperilled State.
All parties combined against the Directory, with
characteristic national vehemence, in the panic
caused by defeat and fear; and two of the Directors
were thrown out as a sacrifice, though the change
could produce no good consequence. Meanwhile,
the beaten chiefs of the armies, who for some time
had chafed a good deal against a despised civilian
rule, exhaled their grievances in angry complaints;
and popular leaders, appearing once more, clamoured
for the energy of 1793, and compelled the government to have recourse to laws of an extreme kind
against priests and *émigrés*, and to arbitrary military
and financial experiments. At the same time, at

Marginal notes:

The reverses of 1799 cause all parties to combine against the Directory.

Weakness and ruin of the State.

the news of the success of the allies, La Vendée showed symptoms of rising ; the sources of revenue quickly dried up; the armies, driven upon the frontier, from the fertile tracts on which they had lived, were reduced to a state of extreme want ; and between dread of a counter-revolution, and of a revival of the Reign of Terror, the thoughts of all the moderate part of the nation turned eagerly to what had long been their wish—a strong government that would defend France, and save the interests pro-

Desire for a strong Government.

duced by the Revolution. In the shipwreck which menaced the sinking Republic, the ominous words

Siéyès.

'we must have a chief' dropped from Siéyès, the most far-sighted of the governing Five ; and in the Legislature, the armies and the great body of the people, a sentiment which had been growing up that a complete change of system was needed acquired at once irresistible force.

Fortunes of Bonaparte in Egypt.

While this was the state of France and Europe, Bonaparte, undismayed by dangers around, had been

---

¹ The Abbé Siéyès was born in 1748, and in 1784 was made Vicar-General of the diocese of Chartres. He devoted himself to political speculation, and having written a pamphlet on the state of the Commons in France, which became famous, was returned to the States-General in 1789. His courage, however, was not equal to his intellect, and he sank into nothingness during the most stormy times of the Revolution. Having joined the party which overthrew Robespierre, he afterwards became one of the Directory, and promoted the Revolution of the 18th Brumaire. He sank into inglorious wealth and repose during the Empire, and lived to see Louis XVIII. restored to the throne.

carrying on his daring enterprise in the corner of
Africa where he seemed imprisoned. Having, in
some measure, pacified Egypt by a policy of mingled
craft and rigour, he advanced into Syria across the
isthmus, not impossibly—such was the wide sweep of
that dazzling yet capacious intellect—with an
ulterior design of reaching Persia and descending
on India by the Euphrates. He was, however,
baffled by English energy in an attempt to secure a
hold on the coast; and having, to his bitter dis-
appointment, raised the siege of Acre, he was forced He fails at
to retrace his steps to Egypt. He was before long Acre, March to
assailed by the Turkish hordes sent by the Porte to May, 1799.
assure his overthrow; but he defeated them with
terrific carnage; and having reached the seaboard,
not far from the spot where he had disembarked
more than twelve months before, he received intelligence for the first time of the great reverses of
1799. His resolution was taken at once; and if
ambition was his ruling motive, it is puerile to
charge him with fear and perfidy. He gave his On hearing
command to Kleber, his well-tried lieutenant, his the news of the state of
army being at the moment safe, and even without France, he leaves
an enemy at hand; and he set off without delay for Egypt.
France, where, he rightly conjectured, his presence
was sought, and where, too, such a man was wanting
He landed in October, 1799, on the shores of Provence, having fortunately slipped through the English fleets; and his landing, when known, became

the signal for a burst of national and heartfelt welcome which revealed the instincts of the great mass

of Frenchmen.   At every stage on his way to Paris
he was greeted by enthusiastic crowds, as the last
hope of France in her hour of misfortune ; and the
feelings of the soldiery rose to the height of fanaticism at the sight of their well-known leader.   In
the capital the excitement was as intense ; the populace and the garrison openly hailed the conqueror of
Italy as the Chief of the State ; and even the Councils and the Directory, swept along by the vehement
tide of opinion, felt or feigned reverence and exultation.

In this state of affairs the existing government
could not continue for any length of time.   Within
a few days Bonaparte had become the real centre of
political power ; all parties, except the extreme Republicans, who instinctively felt he was a deadly
enemy, and especially the new aristocracy of riches,
gathered round him with anxiety and hope ; and the
chiefs of the Army readily concurred, although divided by mutual jealousies.   Two of the Directors,
Siéyès being one, belonging to the enlightened
Moderates, assented to the Revolution visibly impending ; and a majority of the Ancients agreed to

second another *coup d'état* in the interest of Bonaparte.   That leader, who with his wonted insight
had seemed to keep aloof, and had bided his time,
now made his preparations for the crisis at hand ;

and if his acts were marked by stratagem and guile, they were not stained by the cruelty and blood which had hitherto been the disgrace of similar changes. On the allegation of a Jacobin plot in Paris, the Ancients voted, on the 18th Brumaire (November 9, 1799), that both the Councils should repair to St. Cloud, the object being to deprive the Legislature of the means of resistance, and to dissolve it quietly. Meanwhile, the garrison of the capital had been gained; watches had been placed on the National Guards, and the Heads of the long powerless Commune, in order to prevent a further outbreak; the habitation of Barras, Gohier, and Moulins, the three Directors, not in the secret, was surrounded by troops; and Siéyès and his colleague Ducos broke up the government by a formal resignation of the offices they held.   Bonaparte was thus made suddenly master of Paris, with the soldiery and its leaders devoted to him; and as all that he had done was welcomed by the immense majority of the citizens, his easy triumph appeared assured.   Of all Powers, however, a popular assembly most keenly resents an act of indignity; and the Council of Five Hundred, when it found itself deceived and decoyed away on a mere pretext, broke out in fierce and threatening complaints, though largely composed of the very party which secretly desired a change in the State.   On the following day, Bonaparte appeared at St. Cloud, 'to explain,' as he said, 'his conduct;' but he was met with ex-

The 18th Brumaire, November 9, 1799.

Scene in
Assembly
at St.
Cloud.
clamations of hatred and terror; and for a moment his position was critical, for the guard around the Assembly wavered. The die, however, had been cast; the president of the Five Hundred, Lucien, a brother of Bonaparte, declared the Council lawfully dissolved; the hall was cleared by armed men of hostile deputies, and a sufficient number remained to sanc-

Formation
of a provi-
sional
govern-
ment.
Bonaparte
First Con-
sul.
tion the already accomplished transfer of power. A provisional government was next appointed; but though it was composed of three consuls, two being Siéyès and his facile colleague, Bonaparte, as First Consul, was really supreme.

Character
of the Re-
volution of
the 18th
Brumaire.
Such was the *coup d'état* of the 18th Brumaire, one of the principal events in the French Revolution, and, indeed, in the annals of modern Europe. The Republic was to exist for a time in name, as the Roman Senate survived Pharsalia; but though the truth was veiled under decent forms, the new Cæsar was everything from the first; and history before long was to repeat itself, and to see the rise of a Cæsarian empire. In overthrowing the existing government of France, Bonaparte, doubtless, acted without scruple, and was not superior to ambitious selfishness; and in the snare he laid for the Five

Reflections
on the con-
duct of
Bonaparte
and on the
march of
events.
Hundred—an unwise deception, which provoked to anger an assembly really not hostile—we see that contempt of popular sentiment, and of everything associated with popular forces, which was a distinctive mark of his character, and one of the most

striking defects in it.   But the Catos who denounce
Brumaire as a crime and an ‘assassination of French
liberty’ simply misunderstand or distort facts; and
views such as these entirely miss the true nature of
the French Revolution. Bonaparte had really France
on his side in thrusting the Directors from their
seats, and merely accelerated the course of events
which had long pointed to military rule ; and History
can fairly say for him that the Dictatorship he seized
was perhaps needed, was certainly the choice of the
French People, and, as he truly boasted, ‘ cost not a
drop of blood.’ As for the ‘ liberty ’ which he has   The *coup*
d'état was
been charged with destroying, it was a mere figment   not a crime.
without real existence ; and he could not have struck
the Republic down had it had a root in the national
heart.   In fact, the Revolution, in its whole course,
had been unfavourable to the growth of true popular
rights and of republican institutions in any real
sense ; and the nature of events and the disposition
of Frenchmen had concurred to produce that despotism of the sword of which Bonaparte was only
the most splendid image.   It would have been a
task of extraordinary difficulty to have founded anything like freedom upon the corruption of the old
Monarchy ; and the Legislation of the National Assembly, and the passions generated in the war that
followed, only led to anarchy and tyranny combined.
As for the Republic, it was the mere offspring of
passion ; and, after the experience of the Reign of

Terror, a reaction against it quickly set in, which, notwithstanding all the Directory could have done, would have proved irresistible in the course of time. Besides, in the actual state of France and Europe, a Republic which required the nurture of peace could hardly have acquired in any case stability; the short-lived Republic which was set up soon yielded to the influence of the sword; and, tried among a people ill suited to it by temperament and historical tradition, it could, perhaps, only end in failure. The proneness of Frenchmen to bow to power and to admire military grandeur and success hastened the Revolution already at hand, when a crisis of national danger appeared; and the Hour, when it came, found a Man who satisfied the wants, the hopes, and the fancy of the Nation. In these circumstances the 18th Brumaire can be hardly matter of surprise or censure, though in the suddenness of the Revolution itself we see another proof of the passionate mobility and changeableness of the French character.

CHAPTER X.

MARENGO. LUNEVILLE. AMIENS.

THE first care of the new ruler of France—the as- *Wise and healing* cendency of Bonaparte was at once complete—was *policy of the First* in some measure to restore the finances, the condi- *Consul.* tion of which had become deplorable. The First Consul had brought to this task a resolute will, a commanding intellect, and a faculty of organisation perhaps never surpassed; and enjoyed advantages, to carry out his object, beyond the reach of the fallen government. The moneyed classes, who had given him support in the Revolution which had placed him in power, advanced readily considerable funds to supply the needs of the exhausted treasury; and, as the resources of France were really immense, trade revived and the revenue increased at the first sign of order and confidence. The government, *Financial* however, had recourse to other means to place the *reforms.* finances on a better footing, and to some extent to improve public credit. Bonaparte obtained the ser- vices of a very able man, who had refused to hold office from the Directory; and under the skilful care of Gaudin—a minister of real capacity and worth—a

series of admirable reforms were begun in the whole
financial system of the State.   The iniquitous forced
tax on the rich was abolished ; and while the direct
taxes which since 1789 had formed the only ordinary
sources of supply were distributed and raised with a
regard to justice, an attempt—feeble indeed, and
tentative at first, but ultimately leading to marked
results—was made to return to some of the indirect
taxes, which had been recklessly and unfairly removed in the transports of revolutionary passion.
At the same time a thorough change was made in
the mode of collecting the public imposts, which
had been wasteful and offensive alike; and, by
arrangements in some degree borrowed from the
practice of the old Monarchy, but modified and improved by modern experience, receipts were rendered
more quick and certain, while considerable sums
were immediately obtained from the new officials
who had become collectors.   Provision was also made
for the payment of the debt, or rather what remained of it, which had been unpaid for a considerable time, and before long national insolvency had
ceased.   The surviving Jacobins and extreme Republicans complained truly that more than one of
these measures had too much in common with the
old order of things; but with the First Consul this
was an idle objection, and these reforms were alike
judicious and able.   Sufficient means were in this
way obtained to satisfy the most pressing wants of

the State, and especially to relieve the armies, the
distress of which had become lamentable ; and the
foundations were laid of financial order. The First
Consul, however, went much further; and, as soon as
he had secured power, he endeavoured to bind up
the wounds of the Nation, and to mitigate the animosities which distracted France. His position and
the accidents of his life contributed largely to serve
his purpose ; for, as authority really centred in his
hands, and he had taken no part in the Revolution,
it had become possible, especially in the comparative
quiescence of the passions of the past, to carry out
a policy at once more equable, more firm, and of a
more conciliatory nature than the Directory or Convention could have attempted. The benefits he
conferred in this respect on France from the first
moment do not admit of question. His attention
was directed to the clergy first, who, as we have
seen, had been an object of jealousy and proscription
for many years, and had been persecuted with fresh
rigour after a brief instant of illusory clemency.
The First Consul, until the time should arrive for a
permanent settlement of ecclesiastical affairs, procured the repeal of the most severe laws of the Revolution against the Catholic priesthood ; and he
weakened a source of sacerdotal hate by substituting
for the irritating test imposed by the National Assembly a simple oath of allegiance to the State, to
be taken by clergymen of all descriptions. This

Fortunate position of Bonaparte as a media- tor between factions.

Laws against clergy and emigrés re- pealed or mitigated

P

wise policy made good subjects of thousands of men who had hitherto used their influence against the whole course of the Revolution since 1789; and as ministers of religion of all kinds were not only tolerated but even encouraged, while perfect freedom of conscience remained, the fierce dissensions lessened to some extent which had been so grievous in this particular. The next soothing measure of the First Consul was to abolish many of the sanguinary decrees indiscriminately passed aganst the *émigrés* in a mass, and to extend an amnesty to certain classes of them; and in this manner a number of exiles who hitherto had fought in the ranks of her foes began to return to France, to support the new government, and to detach themselves from the Bourbon cause. Finally, the troubles which had

Pacifica-
tion of La
Vendée.

arisen in La Vendée were appeased, to a considerable extent, by recurring to the judicious policy of Hoche, carried out with a firm yet clement hand; and though one or two severe examples were made, the system of moderation was the general rule.

Rapid re-
covery of
France.

By these prudent and just measures the State, which seemed in hopeless decline, regained speedily new life and vigour; and France was restored in a few months to an extent which might have been thought impossible. Meanwhile the task of framing a new constitution for the still nominal Republic had been given to Siéyès, the ablest of the makers of systems who had been so numerous in the Révo-

lution. It would be hardly necessary to notice the-
results of this work, which left the greatest changes
of 1789, now beyond recall, entirely intact, and'
merely substituted a new State machinery for that
which had been swept away, if it were not that the
' Constitution of the year VIII.'—this was the name-
of the new political growth—illustrated curiously
the cast of thought now prevalent among men 'of ex--
perience in France, and supplied some of the illusory
forms which concealed the power of the new Chief'
of the State. ·

The real objects of Siéyès were to maintain
popular rule in appearance, and yet to curb the excesses of popular license, and at once to create a strong
government, and to guard against the irregular
tyranny of which he had seen such frightful examples. For this purpose, in strange contrast with
the ideas of not many years before, his scheme preserved an image of popular rights, and declared
that Sovereignty belonged to the people ; but it
confined the whole administration of the State, even
in its lowest departments, to certain lists of citizens,.
and it distributed the Legislative and Executive-
powers between a council of State and a Tribunate,
charged respectively to propose and discuss all
measures, a Legislative Assembly the duty of which
was to enact laws without the agitation of debates,
a Senate to nominate to all great offices, and a Grand
Elector and two Consuls to govern under all kinds

Constitution of the year VIII.

The institutions founded by it.

of restrictions. By these means the ingenious de-
signer imagiñed that he would reconcile democracy
with stability, order, and political freedom; but it
is unnecessary to say that his pretty system found
little favour in the sight of the ruler of France.
Bonaparte allowed the limitations on popular rights
in the choice of functionaries to continue for a time;
and he approved of the silent Legislature and the
divided duties of the Tribunate and the Council of
State; for such an arrangement, he clearly saw,
weakened any influence these bodies could possess.
But he insisted on curtailing the privileges of the
Senate, and in removing checks on the Executive
power; and he placed himself, with the title of First
Consul, and with absolute control over the whole
scheme of government, in the stead of the Grand

Elector and his two dependents. The dictatorship
of the First Consul was to last for ten years, and a
second and third consul, mere shadowy names, were
to veil the reality of a single ruler.

In this way the despotism of one man, surrounded by merely nominal restraints, became definitely established in France, and the supremacy of
Bonaparte was consecrated by law. The Constitution
received the sanction of an enthusiastic popular vote,
significant of the national instincts and character;
and before long Siéyès and his late brother Director
gave way to two new consuls, who, though able men,
were merely the willing agents of power. The

government of France, however, at this juncture
was only a small part of the First Consul's task ; he
had, if possible, to repair the disasters of war, and
to roll back the Coalition from the frontiers. These
cares had, of course, engaged his thoughts as soon as
the reins of power had come into his hands; and,
supported by a strong national sentiment, and by
able and skilful lieutenants, and employing the
growing resources of the State with absolute authority and consummate art, he soon reorganized
the shattered armies and revived the military strength
of France. One circumstance was of happy omen
to him, for the Czar, after the defeat at Zurich, had
ordered Suwarrow to return home ; and thus the
forces of the Allies were reduced by a large contingent of hardy warriors. The German Empire
and England, however, remained in the field; and
while an army of considerable but inferior strength
threatened Alsace from Western Bavaria, Austria
had assembled a very powerful force to secure the
conquests she had made in Italy, and having reduced the Italian fortresses, still garrisoned by weak
French detatchments, to invade the borders of
Dauphiny and Provence. In this state of affairs the
First Consul formed a plan of operations which has
been always thought one of the most dazzling of his
military conceptions. The force of the enemy in
Bavaria was not so great as it ought to have been,
regard being had to the whole theatre of war ; and

Reorgani-
zation of
the French
armies.

Plans of
the First
Consul.

the great Austrian host on the western verge of Italy was dangerously exposed on this secondary frontier, to an attack from Switzerland, which projected, like a huge natural bastion, along its flanks and rear. Bonaparte, accordingly, arrayed a force superior to that in its front in Bavaria, which was to descend rapidly from the heads of the Rhine, and, under Moreau, to take the foe in reverse ; while he prepared secretly a second army, with which, concealing his design to the last moment, he resolved to cross the great Swiss Alps, and to fall on the Austrians and cut off their retreat.

The campaign of 1800.

Operations began on both sides in the spring of 1800 on the theatre of war. Leaving Ott to undertake the siege of Genoa, and covering his line of retreat with a large scattered force, Melas, the Austrian commander-in-chief in Italy, advanced to

Operations of Melas in Italy,

the Var and had soon taken Nice. Meanwhile, Moreau had set his army in motion ; and though too timid to carry out the project of striking his enemy in the rear by crossing the Rhine at its heads

and of Moreau in Bavaria.

at Schaffhausen, he had nevertheless invaded Bavaria, forced back his weaker antagonist Kray, and, as had been agreed on, was able to send a considerable detachment across the St. Gothard, to co-operate with the First Consul. That great commander had in the interval drawn together gradually from all parts of France the troops intended for the decisive stroke ; and he screened the movement with such skill that the Austrians believed it was nothing more than the preparation of a levy of conscripts. By the middle of May 50,000 men, ready for the field, were on the Swiss frontier, and everything had been arranged with clear forethought for overcoming the great barrier before them. Sending a column by the ordinary pass by Mont Cenis to deceive the enemy as long as possible, the First Consul directed the mass of his army over the Great St. Bernard; and from May 16 to May 19, the solitudes of the vast mountain tract echoed to the din and tumult of war as the French soldiery swept over its heights to reach the valley of the Po and the plains of Lombardy. A hill fort, for a time, stopped the daring invaders, but the obstacle was passed by an ingenious stratagem; and before long Bonaparte, exulting in hope, was marching from the verge of Piedmont on Milan, having made a demonstration against Turin, in order to hide his real purpose. By June 2 the whole French army, joined by the reinforcement .sent by Moreau, was in possession of the Lombard capital, and threatened the line of its enemy's retreat, having successfully accomplished the first part of the brilliant design of its great leader.

The First Consul crosses the Alps May 16–19, 1800.

The French army enters Milan, June 2.

While Bonaparte was thus descending from the Alps, the Austrian commander had been pressing forward the siege of Genoa and his operations on the Var. Masséna, however, stubbornly held out in

Genoa; and Suchet had defended the defiles of
Provence with a weak·force with such marked skill
that his adversary had made little progress. When
first informed of the terrible apparition of a hostile
army gathering upon his rear, Melas disbelieved
what he thought impossible; and when he could no
longer discredit what he heard, the movements by
Mont Cenis and against Turin, intended to perplex
him, had made him hesitate. As soon, however, as
the real design of the First Consul was fully revealed,
the brave Austrian chief resolved to force his way

Melas falls
back.

to the Adige at any cost; and, directing Ott to raise
the siege of Genoa, and leaving a subordinate to hold
Suchet in check, he began to draw his divided army
together, in order to make a desperate attack on the
audacious foe upon his line of retreat. Ott, however,
delayed some days to receive the keys of Genoa,
which fell after a defence memorable in the annals
of war; and, as the Austrian forces had been
widely scattered, it was June 12 before fifty thousand men were assembled for an offensive movement round the well-known fortresses of Alessandria.
Meanwhile, the First Consul had broken up from
Milan; and whether ill-informed of his enemy's
operations, or apprehensive that, after the fall of
Genoa, Melas would escape by a march southwards,
he had advanced from a strong position he had
taken between the Ticino, the Adda, and the Po,
and had crossed the Scrivia into the plains of

Marengo, with forces disseminated far too widely.
Melas boldly seized the opportunity to escape from
the weakened meshes of the net thrown round him;
and attacked Bonaparte on the morning of June 14
with a vigour and energy which did him honour.
The battle raged confusedly for several hours; but
the French had begun to give way and fly, when the
arrival of an isolated division on the field, and the
unexpected charge of a small body of horsemen,
suddenly changed defeat into a brilliant victory.
The importance was then seen of the commanding
position of Bonaparte on the rear of his foe; the
Austrian army, its retreat cut off, was obliged to
come to terms after a single reverse; and within a
few days an armistice was signed by which Italy to
the Mincio was restored to the French, and the disasters of 1799 were effaced.

Battle of Marengo, June 14, 1800.

The French recover Italy.

This splendid success has been always considered
one of the most wonderful of its author's achievements. Yet, though the daring passage of the Alps
was a military combination of the highest order,
carried out generally with the greatest skill, the
movements of Bonaparte in this campaign hardly
equalled those of 1796, and the march on Marengo
gives proof of that over-confidence and sanguine
ardour which were the chief defects in his genius for
war. While Italy had been regained at one stroke,
the campaign in Germany had progressed slowly;
and though Moreau was largely superior in force, he

Campaign in Ger- many.

Advance of
Moreau.
Ability of
Kray.
had met more than one check near Ulm, on the Danube. The stand, however, made ably by Kray, could not lessen the effects of Marengo ; and Austria, after that terrible reverse, endeavoured to negotiate with the dreaded conqueror. Bonaparte, however, following out a purpose which he had already made a maxim of policy, and resolved if possible to divide the Coalition, refused to treat with Austria jointly with England, except on conditions known to be futile ; and after a pause of a few weeks hostilities were resumed with increased energy. By this time, however, the French armies had acquired largely preponderating strength ; and while Brune advanced victoriously to the Adige—the First Consul had returned to the seat of government—Moreau in Bavaria marched on the rivers which, descending from the Alps to the Danube, form one of the bulwarks of the Austrian Monarchy. He was attacked incautiously by the Archduke John—the Archduke Charles, who ought to have been in command, was in temporary disgrace at the Court—and soon after-

Battle of
Hohenlin-
den, De-
cember 3,
1800.
wards he won a great battle at Hohenlinden, between the Iser and the Inn, the success of the French being complete and decisive, though the conduct of their chief has not escaped criticism. This last disaster proved overwhelming, and Austria and the States of the Empire were forced to submit to the

Treaty of
Luneville,
February 9,
1801.
terms of Bonaparte. After a brief delay peace was made at Luneville in February 1801 ; and the

glorious provisions of Campe Formio were ratified
and extended in the interests of France.   The principle of the French 'natural boundaries' was confirmed again by a second cession of Belgium and the
western bank of the Rhine ; and the young Republics
created in Italy—they had remained true to their
allegiance to France—were, save at Rome and Naples,
set up again, and recognized by Austria, although
the objects of the bitter aversion she naturally felt
for what represented revolution and defeat.   The
Grand Duke of Tuscany, an Austrian prince, was in
addition deprived of his duchy, to be conferred on
an Infant of Spain, a Power now wholly dependent
on France ; and the First Consul pursued the system
of secularising, as it is called, the great German
Bishoprics, in order to satisfy the greed of Prussia,
to bind her more closely to the French alliance, and
effectually to divide Germany.   The Pope, however,
was recalled to Rome, as the First Consul had need
of his support in a great measure already impending ; and, at the intercession of Russia, Naples was
spared, and hopes were held out to the dethroned
King of Sardinia.   In negotiating this treaty, which
not only assured to France the coveted boundary of
the Rhine, but made her dominant over half the
Continent, Bonaparte had shown the art of the
young General of 1796, and the same contempt of
revolutionary principles.   But he had assumed a
more dictatorial tone, and hardly a trace of the

Great advantages gained by France.

Dictatorial tone of Bonaparte.

moderation remained of which he had given proof
when his power was uncertain. The change was felt,
and it was generally perceived by the representatives
of the old Powers of Europe that the ambition and
craft of the military genius who ruled France might
be at least as dangerous as the propaganda of Republican liberty. England was now again left alone
to contend against the State which had twice defeated Europe; and many circumstances concurred
to make her wish to give up the struggle, at least
for a time. Her ascendency on the sea had become
as evident as that of her antagonist on land. She
had swept the fleets of France from the ocean, and
conquered most of the French and Dutch colonies;
and, in the existing state of the world, the rival
nations could not find a theatre for a decisive
encounter. Mr. Pitt, too, had retired from power;
the continuance of the war had become unpopular,
and even the Tory majority in the two Houses felt
that an interval of repose would be welcome. Events
hastened the consummation to which things had
been of late tending. An English force had landed
in Egypt, and, retrieving years of military discredit,
had compelled the veteran army of Bonaparte to
capitulate after a gallant struggle, and Egypt had
been definitely lost to France. On the other hand,
the First Consul had combined a formidable league
against England, headed by the maritime Powers of
the North; and though this alliance was quickly

March–
August
1801.

dissolved, it caused apprehension not lessened by the
failure of Nelson to destroy a French flotilla off the
coast of Boulogne. After long negotiations peace
was signed at Amiens, in March 1802, France retaining all her continental conquests and recovering
some of her colonies, England keeping Ceylon and
Trinidad; and though statesmen felt that it was
only a truce, the two nations rejoiced that the sword
had been sheathed. One article in the treaty, bitterly
discussed, was soon to become of unhappy importance. Malta had been wrested by our fleets from
the French; and it was stipulated that the great
fortress should be restored to its original possessors,
on the condition, however, perfectly understood,
that France was not to make fresh annexations in
Europe.

March–April 1801.

Treaty of Amiens, March 27, 1802.

In this way the most dreaded enemy of France
retired from a contest of nine years, and the supremacy of her rival on the Continent was confirmed.
The Peace of Amiens was immediately followed by a
general pacification of Europe; and the ruler of
France stood before the world encompassed by a
fresh halo of glory and renown. Within the space
of a year and a half that wonderful man had raised
France from what seemed hopeless prostration and
anarchy, had given her order, allayed her troubles,
revived her strength, and struck down her foes; and
he had consolidated her triumphs by a series of
treaties which made her arbiter of the finest part of

Great results obtained by the First Consul.

Europe. The prospect was magnificent, and appeared
serene; but would the warrior who had gathered in
his hands the stormy forces of the Revolution, pause
in the intoxicating career of victory? Would not
the confusion and change of Europe offer to his
ambition a perilous field; would not the animosities
of defeated Powers come into fierce collision with his
imperious rule and the order of things which he
wished to establish; would not the despotism he was
inaugurating in France, though, perhaps, inevitable
in her present state, accelerate the tendency to
foreign conquest which she had displayed for several
years?

CHAPTER XI.

THE CONSULATE. RENEWAL OF WAR.

DURING the period which followed the Peace of Amiens,[1] the First Consul had leisure to make great changes in the internal government of France, to carry out the policy of reconstruction, and of moderating and reconciling the remains of factions begun after the 18th Brumaire, and at the same time to increase his own domination, and to weaken whatever seemed hostile to it. His measures of re-

Internal Government of the First Consul.

---

[1] M. Thiers, in the *Histoire du Consulat et de l'Empire*, has described in the minutest detail, and with a masterly hand, the internal Government and the foreign and domestic policy of Napoleon. This work should be studied for its copious information; but in peace as in war the brilliant author surrounds his ideal with deceptive splendour. M. Lanfrey, in his *Histoire de Napoleon I.*, has said all that could be said on the other side, and has painted the vices and mischiefs of Napoleon's despotism, and the faults of the Emperor's character, with much skill and power. The *Commentaries*, and especially the *Correspondence* of Napoleon, show what he was as a ruler and administrator as well as a warrior, and the works of MM. Bignon and Fain may also be consulted. The correspondence of Lords Grenville, Wellesley, Sidmouth, and Castlereagh, of Mr. Pitt and Mr. Fox, and the *Memoirs* of Prince Hardenberg, show what Napoleon's régime appeared to English and German statesmen; and Alison's *History*, though disfigured by party views, contains a full account of the Consular and Imperial system of Government.

form and pacification had already been followed by
good results, but many of the institutions of the
country he ruled, and its social frame in some of its
parts, were shattered, distorted, and out of joint,
after the frightful shock of the Revolution; and an
opportunity was afforded him to reconstitute the
polity of France in several respects, to leave a permanent mark on it, and to influence powerfully the
national life, in what he conceived the interests of
the State, and in furtherance certainly of his own
objects.   Bonaparte, as we have already said, in consequence of the antecedents of his career, and of the
firm hold he preserved of power, was in many particulars fitted for this work; his great ability gave
him some qualifications for it, and what was more
important, the circumstances of the time concurred

*The time
favourable
for recon-
structing
society in
France.*

largely to serve his purpose.   The Revolution had
now permanently set free the soil, had removed mischievous restrictions on trade, had relieved Frenchmen of feudal shackles, had secured a general
equality of rights more real than the fanciful Rights
of Man, and had laid the foundations of a material
prosperity which was to become equally brilliant and
solid ; and terrible as its devastations had been, these
blessings were great and were to prove lasting.   But
while in the highest places of government it had
ended only in a long succession of weakness and oppression, for the present closed by a splendid but
weighty despotism of the sword, it had also left con-

siderable parts of the national organization a mere
chaos; and order, tranquillity, and reform were
needed in the administrative system, and the
ecclesiastical and even social arrangements of the
long disturbed and agitated country. At the same
time the gradual subsidence of past animosities had
grown more evident; the desire for repose had
become supreme in the extinction of political hopes
and passions; the one thought of the numerous
classes which had gained advantage from the Revolution was to disregard its ideals and to reap its
benefits under a regimen of settled authority and
law; except a few royalists and extreme anarchists,
all parties accepted facts as they were; and the
Nation, entranced by the spell of glory, and grateful
also for splendid services, looked up to Bonaparte
with unthinking confidence. A great field lay open
to the First Consul, and the soil was ready for his
strong hands to turn.

The attention of Bonaparte was first directed to
the whole internal administration of the State. The
reforms effected in the finances had restored credit
and assured the revenue; and the causes were not
in operation yet which in this respect were to lead
alike to illusory prosperity and real exhaustion. But
almost everywhere else, disorder prevailed; and if
much that the First Consul did has been censured
by able thinkers, many of his creations have obtained
the sanction of national approval and have become

Reforms in
the State.

Q

permanent. One of his chief cares was to accomplish a change in the general judicial system of
France, which the National Assembly had, with
strange unwisdom, exposed to the evils of popular
election, and which now required a thorough amend-

The Judi-
cial system.

ment. The appointment of judges was properly
secured, as in England, to the Executive government; justice was brought nearer home to all
Frenchmen by increasing the number of inferior
judges; and uniformity in the distribution of rights
was obtained by establishing a series of appellate
tribunals, in some degree resembling the old Parliaments, but with a better and modern procedure.
This important reform, in which we again see the
ancient Monarchy imitated and improved, was certainly marred by the institution of special courts
for political offences; but it must be recollected
that a system of this kind had at all times existed
in France, and never was so fearfully abused as dur-

The Code

ing the period of the Revolution. The next great
work of the First Consul was to give the Nation the
one general Code which the National Assembly had
projected, and the Convention had endeavoured to
begin; and in a few months, under his energetic
impulse, the medley of usages and written obser-
vances, confused, uncertain, and huge in bulk, which
had formed the canon of rights in France, were
fused into a harmonious body of laws, the real
merit of which is seen by their extension over a

great part of Europe, though their genius is in many respects despotic. This noble achievement was of course, in the main, the task of professional lawyers; but Bonaparte may claim to have been its chief author, and in some places even the text of the Code bears the mark of his keen and powerful intellect. Perhaps, however, the most notable of the internal changes made at this time was the Revolution which the First Consul wrought in the arrangement of local powers in the State, and the relations of these with the supreme government. The Convention had, as we have seen, limited the extravagant authority given by the National Assembly to local bodies which had proved so mischievous in the Revolution, and Bonaparte carried to the furthest extent the principle of restriction and compression. The powers of the provincial and municipal Assemblies were almost everywhere wholly suppressed; the influence of the Communes, including that of Paris, already curtailed, was reduced to nothingness; the National Guards were made simply a submissive appendage of the Army; and France in local affairs was practically ruled by a bureaucracy of sub-prefects and prefects, in close dependence on the central government, and in many respects with a strong resemblance to the royal intendants of the Bourbon Monarchy.

Centralization of local powers.

Prefects and sub-prefects.

The next great measure of the First Consul was the renewal, under altered conditions, of that alliance

The Concordat.

between the State and the Church, which had existed in France since the dark ages. He had, as we have said, on his advent to power, put an end to the persecution of the clergy; but after the events of past years, confusion prevailed in all parts of the Church, and its relations with the people and the government alike were jarring, irregular, and ill-defined. A bitter feud divided the nonjuring priesthood from those who had taken the oath of 1790–91, and communicated itself to their flocks everywhere; and as the great majority of the bishops were *émigrés*, and many of their Sees had become vacant, whole districts were without the episcopal rule which is an essential part of Roman Catholic discipline. Besides, France had been under a kind of interdict since the property of the Church had been swept away, and the position of its ministers changed by the Legislation of the National Assembly; and the open disfavour of the Holy See had added to ecclesiastical disorders, been a source of real weakness to the State, and shocked the consciences of millions of Frenchmen. In this state of things the First Consul, after long negotiations with the Papal Court, at present really well-inclined to him, obtained what was called a Concordat with Rome, the first of many arrangements of the kind, which in some measure at least reconciled the ecclesiastical and civil powers in France, allayed many of the troubles of the Church, and at once bound it up with the new order of affairs

and placed it under the control of the government.
By this famous compromise, complete freedom to all
sects were permanently assured; the confiscations of
the Church lands were confirmed; the number of
Sees in France was greatly reduced, and their occu-
pants, with the whole body of clergy, were made
simply pensioners of the State; and the supreme
authority of the civil ruler in ecclesiastical matters
was solemnly asserted. But on the other hand the
Catholic religion was declared that of the Nation as
a whole, its organization was upheld by law, and its
teachers given a recognized rank; and if the Church
lost finally its ancient pretensions, and was associated with a Revolutionary State, its internal condition was rendered secure by the support and
favour accorded to it, and the strife within it was
greatly lessened by the complete equality with which
its Ministers, whatever their antecedents, were always
treated. The re-establishment of the Church in
France, and its restoration to a place in the State,
were celebrated by religious ceremonies at which
Bonaparte assisted in person; and—strange spectacle
in that age of wonders—the aisles of Notre Dame,
where a few years before the Goddess of Reason, in
the midst of Revolutionary worshippers, held her
orgies, echoed once more to the sacred services of
the most mystic form of the Christian faith at the
bidding of a revolutionary soldier.

It may be doubted whether the Concordat has Its effects.

ultimately advanced religion in France, while it has
placed the Church and all spiritual affairs in complete subjection to the secular arm.   It tended, however, to restore order, to allay discord, and to promote
peace; and if it enlarged the influence of the new
ruler, it is too much to say that he had no other

Public In-
struction.

motive.   The Concordat was followed by a scheme
of public instruction which also extended the power
of the government over the Nation by putting intellect under the control of the State; but here also
it is unfair to assert that this was the sole object of

General re-
sults of
these re-
forms.

the First Consul.   The general effect of the various
measures of which we have faintly traced the outline
was to improve extremely the administration of
affairs, to diffuse tranquillity, and upon the whole to
contribute to the national welfare; and if some of
these reforms, especially the return to centralization
in local government, have had evil results of their
own, and if they certainly made despotism more
widespread and universally felt, they have almost all
survived to this day, and have satisfied the wants of
the great body of Frenchmen.   The First Consul,
however, took other means to consolidate his sway

Changes in
the army.

which we must briefly notice.   His authority rested
ultimately of course on the instrument by which he
had attained power, and he not only improved the
discipline and organization of the French Army, but
in a great measure transformed its spirit, overcame
the jealousies of its chiefs which had made them-

selves more than ever apparent, and converted into
enthusiastic 'devotion to himself its National and
Revolutionary instincts. But like all soldiers who
have displayed capacity as rulers for political affairs,
he concealed the omnipotence of the sword in the
State, and he endeavoured to obtain support for his
government from the elements of civil life in the
Nation by associating it with a great mass of interests created by, and dependent on, himself. For
this purpose, while he generally maintained the
equality of Frenchmen before the law, he gradually
formed out of the official classes, which he multiplied in every conceivable way, an aristocracy of a
new type; and he tried by every means in his power
to amalgamate it with whatever remained of the
aristocracy which the Revolution had spared. To
accomplish this object he introduced again all kinds
of distinctions in social life—the Legion of Honour
was the most remarkable—and at last ventured on
restoring titles; and by these means he no doubt
strengthened the attachment to himself of the upper
orders of Frenchmen who had risen to influence
through the Revolution, and even allied them to
some extent with the thinned relics of the old Noblesse; though this new patriciate, as has always
happened in similar cases, was a poor creation, unstable, untrustworthy, and little respected. Before
long he effected another great change, which indicated whither events were tending. The Consulate

Creation of a new aris-tocracy.

The Legion of honour. Restoration of Titles.

Bonaparte made Consul for life. for ten years became one for life; the authority of the Senate, composed of the creatures of the new ruler, was at once augmented and made more to depend on his will; and while the restrictions on the popular voice invented by Siéyès were nominally lessened, the Tribunate which had offered a show of opposition on several occasions to the autocrat in power, was reduced in numbers, carefully weeded, and practically merged in the mute Legislative Body.

Modification in a despotic sense of the Constitution of the year VIII. This ' reform of the constitution ' of the year VIII., which made Cæsarian despotism perfect, was, as before, sanctioned by a general vote of an overwhelming majority of Frenchmen.

Partial Resemblance of the new Government to that of the Monarchy. By these means the government of Bonaparte became essentially the domination of one man, well ordered, spreading itself everywhere, gathering to itself all the forces in the State, shaping and controlling the national life, and surrounded by a gradation of powers and a set of influences which gave it a support. A considerable part of the new mechanism of the State had much in common with the ancient system which the National Assembly had tried to destroy for ever; but though a certain resemblance existed, the rule of Bonaparte, in most essential points, differed widely from that of the Bourbon Monarchy; for if it was more despotic and sometimes oppressive, it was more national, and on the whole just. Its great and fatal evils of course

Its evils. were that it left everything to the will of one man,

that it was wholly inconsistent with anything like
liberty, that it more or less weakened the national
energies, even when apparently most beneficent;
that it was, at best, attended with precarious good,
and might issue in grievous mischiefs; that, in a
word, even in its most brilliant form, it was despotism
with all the resulting perils. The Dictatorship, Its merits.
however, of the First Consul, vicious though it was
as a scheme of government and destined to end in
terrible misfortune, had nevertheless the real excel-
lences that it secured internal quiet to the State and
gave France a variety of institutions which have
stood the infallible test of time, and that it alike
protected the order of things which had grown up
under the Revolution, and reconciled it in some
degree with the past; and it is for this, among
other reasons, that it remains dear to the memory of
Frenchmen. As the administration of Bonaparte at Wise Ad-
home was generally at this period moderate, the tion of the
benefits that followed were almost unmixed. He, sul.
indeed, showed himself implacably severe to the remaining dregs of the Jacobin faction; and more
than once treated with unsparing tyranny those
whom he called the 'men of the September massacres.' But he went on steadily with the auspicious
work of reconciling and moderating parties; and he
finally closed the list of the *émigrés*, and admitted
numbers of the exiles into the service of the State.
At the same time the noble system of public works

which have illustrated his era was set on foot; the
canals and roads which had been the pride of the
ancient Monarchy, and for many years had been in
a state of decay, were restored; and new towns
springing up in La Vendée, the capital adorned with
magnificent buildings, and the Alps spanned by vast
military lines, attested the energy of the chief to
whom France had committed her fate.

He encour-
ages the
movement
towards
Monarchy.

Meanwhile Bonaparte already wearing, as it was
said, the 'the shadow of a kingly crown,' promoted
carefully, by indirect means, the domination he had
directly established, and hastened the movement towards monarchy which had been visible even before
his time. He abolished the ceremonies in which the
Republic commemorated the execution of Louis
XVI., and caused the remains of Turenne—the
great hero warrior of the most glorious days of
Louis XIV., which even Jacobin frenzy had spared,
though during the excesses of the Reign of Terror it
had desecrated the rest of the Bourbon kings—to be
transported solemnly to the Invalides, and buried
with extraordinary pomp. He had already taken up
his abode at the Tuileries, and effaced the marks
which revolutionary passion or republican frenzy had
left on the spot; and he held what really was a
Court, with its accessories of etiquette and splendour,
in that seat of fallen yet not forgotten royalty. At
the same time he adopted a regal style in his correspondence with foreign Powers; and though in his

relations with the Bodies of the State he preserved
forms of simple equality, and spoke and bore himself as a private citizen, he always appeared in Paris
with a magnificent retinue, and flattered the populace with the display of grandeur. He also encouraged in every way the luxury and taste of the
Bourbon days, and spoke with contempt to those in
his confidence of the savageness of revolutionary
manners and of the absurdity of republican ways;
and in his serious moments he would often dwell on
the instability of the institutions of France, on the
necessity of settled power in an old State, on the
evil effects of the philosophic theories—the ideology
he scornfully called them—which had swayed the
minds of men a few years previously. Nor were the
tendencies of which he set an example less clearly
apparent in the tone of general opinion, practice,
and sentiment. France teemed with addresses
shedding incense on ' the new Saviour of social
order'; and the Press, lately so anarchic and wild,
but now controlled by a watchful police, poured
forth homage in floods to greet the ruler who had
' closed the terrible age of Revolution.' In the same Change of
way the mimicry of Republican tastes which had manners in
France.
been the mode a short time before disappeared in
the salons of the capital; the cant of classical liberty
was heard no more; ladies put off the Ionic costume
of the Aspasias and Phrynes, of Greek times; and
military brilliancy, costly liveries, and the graces,

the finery, and the frivolity of Versailles, showed themselves again in the new masquerade in which the high life of Paris and France figured. The First Consul had literally become the 'mould of form' for nine-tenths of Frenchmen, and all France yielded to the spell of his influence.

Foreign Policy of the First Consul.

While Bonaparte had been thus extending his sway, and reorganizing and transforming France, he had been not less active and stirring abroad. In his foreign relations at this period he pursued the policy of craft and interest inconsistent with the ideas of 1789, which had distinguished his earliest efforts, and he displayed an imperious will and grasping ambition ; but if he gave proof of that lust for power and domination which was to end in ruin, it should be recollected that the circumstances of the time, and even the conduct of some foreign Powers, contributed to place him in the position he assumed. As if to show his contempt of Republican dreams, he made the Infant of Spain, whom he had chosen for the purpose, King of the ceded Grand Duchy of Tuscany; and in this manner he riveted a yoke already becoming difficult to bear, on the abject necks

Its craft and ambition.

of the Spanish Bourbons. At the same time he annexed Piedmont to France, on the plea that the King had given up the throne; and though he checked revolutionary ideas in Italy, he made the Pope and King of Naples feel that they held their possessions at his will and pleasure. Meanwhile

he increased the hold of France on her new conquests and dependent Republics; and, as might have
been expected, he fashioned the offspring to the submission to himself which the parent displayed. As
President he ruled the Cisalpine Republic, considerably enlarged by the treaty of Luneville, and
given the general name of the Italian· nation; he
managed Holland through a Constitution on the
model of that existing in France, and though he left
Switzerland nominally free, he practically controlled
it as a French Province. The most remarkable ad- French intervention
vance in his power, however, arose from his inter- in Germany.
vention in Germany, due mainly to the quarrels of
German potentates. The secularisation of the Bishoprics, which had been a principle of the Peace of
Luneville, led to angry contentions between the
German Courts, each eager for a greater share of the
spoil; and Austria and Prussia made such exorbitant
demands that the Sovereigns of the lesser States
applied to the powerful ruler of France for aid.
The First Consul gladly became a mediator, and
secured a considerable. increase of territory to
Bavaria, Baden, and Würtemberg, while, to the
extreme satisfaction of Prussian statesmen, he still
further enlarged the bounds of Prussia with the view
of strengthening her against Austria, thus following
traditional French policy which he had made in a
special way his own. In this manner the influence
of France, great in Germany since the day of

Richelieu, was increased in an extraordinary degree ; but if the policy of Bonaparte was hard and calculating, and Germans learned to lament the results, they might bear in mind who called in a protector when they indulge in homilies on French aggression.

Great extension of French power and influence. In this way, in the midst of apparent peace, the domination of France was extended, and her ruler became the undisputed arbiter of Europe from the Baltic to the Mediterranean. It was impossible but that the growth of this power should vex and alarm the only State which had as yet contended successfully with France ; and English statesmen, who had perceived from the first that the peace of Amiens could not last long, began to apprehend that war was near. This was not the case, it was universally felt, of a Republic weak and distracted at home, though strong in its arms and ideas abroad, the existence of which was always precarious ; it was that of a gigantic Despotism, directed and swayed by commanding genius, and all-powerful in France as well as in Europe; and Whigs and Tories equally agreed that the present state of the Continent was a danger to England. In this condition of feeling causes of dissension arose quickly between the two Powers: English politicians not unjustly complained of the

Disputes with England. March to May, 1803. enormous extension of French power and influence, and Bonaparte retaliated by denouncing the asylum given to conspirators against his rule in England, and the hostility of the English Press, of which the freedom shocked his despotic instincts. Meanwhile, on the ground of the virtual infraction of the treaty of Amiens by French ambition, Malta was not ceded at the time arranged; and recriminations on this subject ended in a scene of violence in which the First Consul, breaking out into a real or feigned passion, spoke menacingly to our envoy in Paris. The publication of a French State paper, revealing a design of regaining Egypt, increased the quickening elements of discord, and a kind of challenge which Bonaparte, with his usual scorn of popular forces, threw out generally to the English Nation, aroused an indignation impossible to allay. After fruitless negotiations touching Malta, which, though a principal occasion of the strife, had become merely an incident in it, war was renewed between the two countries; and in May, 1803, the great Powers of the West had again closed in mortal encounter. It is vain to measure the provocation on either side, though in view of the recent aggrandizement of France, the retention of Malta was not contrary to the real spirit of the treaty of Amiens, and though in defying the free opinion of England, the First Consul made a signal mistake which illustrates one of his chief defects as a politician. But if the war was, perhaps, inevitable—for the preponderance of France was perilous in the extreme to England, and this justifies the acts of our statesmen—the renewal of the contest was to be deplored. It was to end in

Renewal of war with England May 18, 1803.

frightful misfortune to France, after raising her to the summit of glory; it was to give England imperishable renown, indeed, and yet to expose her to terrible dangers, to retard her social progress for years, and to involve her in a system of politics with which her people could have no sympathy.

CHAPTER XII.

THE EMPIRE TO TILSIT.

THE new war between England and France was The First the idea of invading our shores, which he had con- embittered by passion, and a death-struggle from plans a sidered premature a few years before ; and he applied descent on the first. The First Consul now took up with ardour England. the idea of invading our shores, which he had con- sidered premature a few years before ; and he applied for months his commanding intellect to preparing the means of a formidable descent. Times had changed since he had advised the Directory to pause, and not to run the risk of the enterprise; he had absolute control of the naval resources of France, Holland, and Italy, largely increased, with those of Spain soon to be added to them ; his military forces overawed Europe, and nothing seemed too difficult for the daring warrior who had hardly met a check in his career of triumphs. Within a short time an immense flotilla, comprising more than two thousand boats, and light vessels with powerful guns, had been constructed along the seaboard extending from La Rochelle to Antwerp; and by degrees this menacing array was drawn together to the coast of

Picardy, and, under the protection of miles of batteries, was collected in the narrowest part of the Channel, within sight of the white cliffs of Dover.

The flotilla and camp of Boulogne.

Meanwhile troops had been marched in thousands from all parts of the dominions of France; and before long the country from Dunkirk to Etaples bristled with the camps of the warlike masses which had been marshalled for the great expedition. Boulogne, and the small adjoining ports, were chosen as the places of embarkation; and the arrangements of Bonaparte were so well laid that his whole army, with its vast material, could be moved on board in a few hours, and the flotilla could be made ready for sea within the space of a single tide. It was not the purpose, however, of this great commander to expose this armament, formidable as it appeared, without ample protection, to the English fleets; and to accomplish this object he matured designs which have always ranked among

Project of covering the descent by a large fleet in the Channel.

his ablest projects. He calculated that the English Admiralty, imagining that the descent would be tried with the powerfully armed flotilla alone, would guard the Channel chiefly with small vessels; and if so, it might become feasible, notwithstanding the naval strength of England, to bring a great fleet into the narrow seas, and under its cover at the decisive point, to effect in safety the dangerous passage. For this purpose he planned a variety of schemes to draw away our squadrons from the waters of Europe,

and to concentrate an armada of fifty sail of the·
line in the straits that divide the two countries;·
and though his combinations ultimately failed, they
were more nearly successful than is commonly supposed.

While Bonaparte was thus straining every nerve
to master what he called 'the wet ditch' of the
Channel, a lamentable incident occurred which has
left a deep stain on his public life, and was ultimately attended with eventful results. The First
Consul had, we have seen, shown generous clemency
to the *émigrés*, and most of them had returned to
France, and even largely entered his service. A
certain number, however, had remained in exile;
and a part of these men, associated with one or two
chiefs of the late western insurgents, had joined in
conspiring against the ruler whose power it was
hopeless to shake openly. As far back as 1801 an
attempt had been made against the life of Bonaparte,.
by firing what was called an infernal machine, as·
he was proceeding to the opera; and this was undoubtedly a royalist plot, though attributed at first
by its intended victim to the survivors of the·
anarchist faction. These machinations, which had·
never ceased, became more active when the war
again broke out, and a project to assassinate the·
First Consul and to destroy his government was
formed in England, though it is unnecessary to·
notice the monstrous charge that English statesmen

Conspiracy
of the *émi-*
grés
against the
First Con-
sul.

connived at it. The Count of Artois, to his lasting
discredit, was cognizant of this criminal purpose,
and it is said, even took part in it; and though the
leaders were men who had fought in the ranks of
the Breton royalists, Pichegru, who had been exiled
since Fructidor 18, was an accomplice to a certain extent; and Moreau, who had become hostile
to Bonaparte, unwisely listened to the tempter's
voice, though innocent of any murderous intent.
The heads of the conspiracy, with Pichegru and
Moreau, were arrested in Paris before they could
effect their purpose; and one of the prisoners having
deposed that a Bourbon prince was to join in the
enterprise, the attention of Bonaparte was unhappily
turned to the Duke of Enghien, a scion of the race,
whose presence on the borders of the Black Forest
had, with other circumstances, aroused suspicion.
The unfortunate prince was suddenly arrested, though
on German territory, and hurried to Paris; and
though guiltless of all real crime, was shot by the
sentence of a military commission, after a trial
which does not deserve the name. Some of the
conspirators were afterwards justly executed; and
the tragedy was closed by the banishment of
Moreau, and by the suicide of Pichegru[1] in his place
of confinement.

 The death of the Duke of Enghien was a crime
which shows what despotism could effect in France,

[1] There seem to be no grounds for the charge that Pichegru was
strangled in prison by the order of the First Consul.

though largely entitled to national gratitude, and seldom marked by mere vulgar cruelty. It is, however, unfair to regard this act as an assassination of the worst kind, for there were grounds to suspect the Bourbon princes; allowance must be made for that dread of murder which has unhinged even the most powerful intellects; and Bonaparte had a right to make an example of the *émigrés* who wickedly sought his life, though he unfortunately selected an innocent victim. The deed, moreover, was less culpable than the slaughter of the French envoys at Rastadt; and if it can be only at best palliated, it is right to bear in mind that the age had acquired a character of violence and angry passion. The immediate effect of this tragic event was to hasten the movement towards Monarchy to which everything had been inclining. The possibility of the sudden death of Bonaparte, which had been brought before the public mind, caused men to hope that the evil results of his disappearance would be at least lessened if he were at once placed on an hereditary throne; and the sentiments of the Nation made it eager to surround its ruler with the pomp of sovereignty. The First Consul naturally flattered these ideas, but whether from a desire to draw a distinction between the position of the Bourbons and his own, or from a wish for new and peculiar greatness, he refused to accept the title of King. At last he selected the ancient dignity which had come down from the time of Charlemagne; and amidst

This event hastens the movement in favour of Monarchy.

The First
Consul pro-
claimed
Emperor of
the French,
May 18,
1804.
enthusiastic demonstrations of joy he was proclaimed Emperor of the French in May 1804, designing himself Napoleon, by his Christian name, according to the usage of Crowned Heads. The Empire, limited to his descendants, was upheld by digni- taries, in part borrowed from the Germanic model, and in part from that established by the ancient Kings of France ; and its military character was fitly expressed by the appointment of sixteen marshals chosen from among the principal chiefs of the Re- publican armies. At the same time fresh changes were made in the shadowy institutions of the work of Siéyès ; and the senate was enlarged, while the popular tribunate was still further weakened, and at last suppressed. More important, certainly, than this mere shifting of the apparatus of despotic power, was the inauguration of the imperial Court, in which the aristocracy of the new era vied with survivors of the old *noblesse*, in flattery, vanity, and ostentation.

Coronation
of Napo-
leon Dec. 2,
1804.
On December 2, Paris flocked to witness the spectacle of the Coronation. In gratitude to the restorer of the faith in France, the Pope had come from Rome to hallow the pageant, and had departed from the usage which his predecessors had imposed on the haughtiest of the German Emperors. The Pontiff, attended by a procession, in which mitres and crosses were strangely mixed with the sabres and banners of the imperial guard, passed along the

Seine to the ancient Cathedral, raised centuries before by the good St. Louis, and still towering in lofty state above the wrecks of the revolutionary tempest. The walls of Notre Dame were hung with tapestry rich with the golden bees of the new Sovereign ; the dim light, which showed nave and aisle, fell on the ranks of the Bodies of the State, of the representatives of foreign Powers, of deputations from the chief towns of the Empire, all arrayed in costly and orderly pomp ; and, as the sacred procession entered, choir and organ pealed forth a solemn chant, and the prelates of the renovated Church of France knelt reverently to implore the apostolic blessing. Meanwhile Napoleon had left the Tuileries, escorted by the new great officers of State, and with the company of his marshals by his side ; and as he moved slowly along the ways which had seen all that was worst in the Reign of Terror, cheers burst exultingly from the thronging crowds, hailing a master as they had hailed liberty. On the arrival of the Emperor the assemblage in the Church stood up to greet him, amidst the swell of sacred music and the blare of trumpets ; and it was with sentiments of mingled curiosity and awe that the spectators beheld the conquering soldier, wearing the golden laurel of the Cæsars on his brow, do homage to the successor of the Galilean fisherman. The ceremony now began, and Pius VII. poured the mystic oil on the kneeling Sovereign, and invested

him with the lesser emblems of power, the consecrated Sword, and imperial Sceptre; but, as he
was about to complete the rite, Napoleon took the
Crown from the hand of the Pontiff, and, with a
significant gesture, placed it on his head himself, in
witness of the supremacy of the State, and of his
own paramount and chief authority. The Emperor
then ascended a throne, encircled by a following in
which great names of the Bourbon Monarchy stood
by the side of republican soldiers and politicians;
and as the hymn arose which had fallen on the ear
of Charlemagne when saluted Emperor of the West,
the acclamations that echoed from Notre Dame were
caught up by the vast crowds outside, and the roar
of artillery joined in concert. The satirist may
ridicule whatever was incongruous or out of date in
the spectacle, but History notes its more suggestive
features—how the Revolution, in Napoleon's hands,
arrayed itself in the forms of the Past, did external
reverence at least to the symbols of majesty, order,
and antique tradition, and embodied itself, so to
speak, in the type of contented servitude and military
despotism.

New coali-
tion against
France.

Before long, however, pageants of this kind gave
way to the sterner scenes of war renewed over the
greater part of the Continent. The execution of the
Duke of Enghien, and the violation of the territory
of a German State, had given natural offence to the
Powers of Europe; and fresh causes of irritation

arose, when, by a transformation expressive of his power, Napoleon converted the Italian Republic into a vassal Monarchy ruled by himself, and incorporated Genoa into the French Empire. Mr. Pitt, too, had returned to office ; and his efforts, in the increasing peril of England, to reunite a confederacy against her foe, soon shaped alarm into definite purpose, and revived the Coalition ever ready to combine. By the summer of 1805 England, Austria, Russia, Sweden, and Naples, had entered into a close alliance ; and it was hoped that even Prussia would join the League, as the overwhelming preponderance of France had begun to affect the policy of that Power, and to make it apprehensive for its own safety. Four lines of invasion were marked out by those who directed the Allied councils ; the first by the North German seaboard, the second up the valley of the Danube, the third from the Adige into Italy, and the fourth along the Neapolitan coast ; but the second attack only was to be made in force ; and the Austrians and Russians who were to attempt it were separated from each other by the immense distance between Bavaria and the Galician frontier. These faulty dispositions were not lost on the great soldier who had so often triumphed over disunited and ill-led enemies ; and Napoleon prepared to defeat the Allies by operations worthy of his genius for war. Comparatively neglecting the subordinate attacks, he resolved to meet the second in irresistible

Plan of the attack of the Allies.

Campaign of 1805.

Napoleon quits Bou-

logne, and
surrounds
and cap-
tures an
Austrian
army at
Ulm, Oct.
19, 1805.

strength, and to crush the Austrians before the
Russians could aid them; and as soon as he had
ascertained that his lingering fleets could not reach
the Channel to cover the descent, he broke up with
his whole army from Boulogne, and marched with
extraordinary speed to the Rhine, while powerful
detachments from Holland and Hanover descended
on the Maine to join in the movement. By the
second week of October these converging masses,
directed with admirable precision and skill, had
gathered on the rear of the Austrian army, which
had been imprudently advanced on Ulm; and within
a few days an iron net was thrown round the doomed
and baffled host, and it was forced to surrender with
Mack its chief. The whole vanguard of the Allied
armies had been thus annihilated by a simple
manœuvre resembling that which had destroyed
Melas; and Europe never witnessed such a scene
again until it was reproduced in our own days by
the capitulations of Metz and Sedan.

Battle of
Trafalgar
and de-
struction of
the French
and Span-
ish fleets,
Oct. 21,
1805.

This wonderful success was soon, however, to be
chequered by a tremendous disaster on the element
on which all the efforts of France were destined only
to end in failure. We have referred to the combinations by which Napoleon endeavoured to collect a
fleet of overwhelming force in the Channel; and
these became in the highest degree formidable,
when, in the autumn of 1804, Spain placed her
naval forces in his hands. In the spring of the

succeeding year a large French squadron set sail from Toulon, and, rallying a Spanish squadron at Cadiz, arrived safely in the West Indian seas, its object being to attract Nelson from European or English waters, and then, joined by a squadron from Brest, to make as quickly as possible its way to Boulogne, and so cover the projected descent. The first part of the scheme completely succeeded; Nelson was led away in a fictitious chase; the French Admiral Villeneuve left the West Indies with a long start over his dreaded rival; and though he was not met by the Brest fleet, he could have hardly been stopped had he made directly for the Channel, which, as Napoleon calculated, was but ill-guarded. But Villeneuve was timid, and inclined southward; a light vessel detached by Nelson, with admirable forethought, gave the alarm; an indecisive action, off the coast of Spain, induced the Frenchman to bear up for Ferrol; and though he had still not a few chances of success, for he had been strengthened by another squadron, he shrunk from his foes, and put into Cadiz. Within a few weeks his whole fleet was destroyed in the greatest naval battle of modern times; and this crushing victory, though dearly bought by the death of the greatest of English seamen, brought all further attempts of invasion to an end. Yet the glory of Trafalgar ought to blind no one to the imminent peril which England escaped; Napoleon's manœuvres

The project of the descent might have succeeded.

were nearly successful; and had Villeneuve had a ray of the genius of Nelson, he would, in all probability, have made the descent possible. What saved England was not the defence of the Channel, which was left too feebly guarded, but the terror of her fleets, and the demoralisation of her foes; and though Napoleon ought to have taken these moral elements more fully into account, he was not far from accomplishing his design. We believe, however, that he entirely underrated the resistance which he would have had to encounter had he succeeded in making the descent; the English army was of considerable strength; and on this, as on other occasions, he unduly disregarded the enormous power of national forces under certain conditions. He might have captured London, but he would, we think, have been ultimately imprisoned within his conquest.

Napoleon marches on Vienna.

Trafalgar, however, was soon forgotten in the exultation of a career of victories. The disaster of Ulm put an end to the scheme of invasion formed by the Coalition; and, having sent detachments to subdue the Tyrol, Napoleon, with the mass of his forces, marched down the Danube on the Austrian capital. The army he commanded was the finest which France, perhaps, ever sent into the field; it had been trained in its camps at Boulogne to the highest point of endurance and vigour; it had been organised upon the system of *corps d'armée*, and separate reserves, since adopted by all Continental armies; though it numbered several German contingents, it was not filled with unwilling auxiliaries, as became the case in subsequent campaigns; and if it had suffered greatly in its late forced marches, it presented a combination of freedom of movement, of activity, energy, and trustworthy force, which justified the name of the Grand Army, thenceforward given it by its mighty leader. The conquering host rolled swiftly onwards, a few Austrian divisions and the Russian army, which had reached the Inn, falling back before it; and after passing the undefended lines of the feeders of the great Austrian stream, it was in possession of Vienna at the middle of November. Meanwhile the Russians, led by Kutusoff, a captain destined to future renown, had judiciously retreated into Moravia, opposing, as obviously was the course of prudence, time and distance to the far advancing enemy; and before long they were encamped round Olmutz, supported by several Austrian detachments. Napoleon, however, having become master of the bridges of Vienna by a stratagem, crossed boldly to the northern bank of the Danube, carrying out his system of daring movements, and relying on the ascendency of immense success; and towards the close of November the Grand Army was collected, apparently in a disseminated state, but really within the hands of its chief, in Lower Moravia, around Brünn and Austerlitz. His position had now become critical, for Prussia,

[marginal note] The Grand army.

[marginal note] Vienna occupied Nov. 13, 1805.

terrified at recent events, had begun to arm, and was about to descend through Bohemia on the French line of retreat, and the Archduke Charles, with his brother John, was hastening with a considerable force from Hungary; and had the Allies simply awaited events, Napoleon must have retired before them. The Czar, however, Alexander, against the advice of Kutusoff, resolved to attack the French Emperor— that great captain had purposely assumed a timid attitude to deceive his foe—and in the last days of November, the Allied forces broke up from Olmütz,

Battle of
Austerlitz,
Dec. 2,
1805.
Ruin of the
Allied
army.

and marched on Austerlitz. An ambitious attempt to outflank Napoleon, and intercept his retreat on Vienna, unduly weakened the line of his enemy; he seized an opportunity which he had foreseen; and, after a fierce and murderous struggle, the Allied army was pierced in the centre, and became a mass of shattered and ruined fragments. The Sun of Austerlitz, to which the conqueror was wont to refer with just pride, saw the warlike strength of the Coalition struck down.

Peace of
Presburg,
Dec. 15,
1805.

This great victory—the masterpiece of Napoleon's tactics on the field of battle—was followed in a few days by a peace, made at Presburg. The Czar lost nothing but military fame; but Austria was compelled to surrender Venice, annexed to the new Italian kingdom; and she ceded the Tyrol to Bavaria, and recognised the Elector as an independent Sovereign. Baden and Würtemburg were also en-

larged, and the Elector of Würtemburg made, too, a Changes effected by it.
King; and thus Austria, the old rival of France,
was reduced to a Power of the second order, and the
policy was carried on of extending the influence of
France among the minor States of Germany. The
King of Naples was soon afterwards dethroned, as a
member of the late Coalition ; and the Emperor of Austria ceases to be Head of the German Empire.
Austria, with a just sense of dignity, acknowledged
his position, and abandoned his claims to the German Empire, long an appanage of his House.
Bavaria, Baden, and Würtemburg, with some lesser
States, were now formed by Napoleon into what he The Confederation of the Rhine.
called the Confederation of the Rhine; and those
German Powers which, in the late campaign, had
proved useful and willing allies, became mere vassals of the French Empire, with their military forces
in the hands of his chief. In this state of things
Prussia was left wholly isolated ; and she was soon
to reap the fruits of a policy which, beginning in Isolation of Prussia.
aggression, had ended in greed. Partly from alarm,
and partly owing to an alleged violation of her territory by the French, Prussia had, we have seen,
prepared to attack Napoleon, when dangerously exposed, in the rear; but after Austerlitz, her govern-
ment recurred to its former course, and had accepted Conduct of that power.
Hanover, for some time occupied by the French
armies, as the price of a renewed alliance with
France, though this perfidy was justly condemned by
her people, and could only provoke the scorn of

Napoleon. The spoliation of the patrimony of the Crown caused England at once to declare war against Prussia; and that Power, having endeavoured secretly to form a new Coalition against France, and a chance of peace with England having arisen on the accession of Mr. Fox to power, Napoleon dealt with Prussia after her own measure, and offered to make over Hanover to Great Britain. This, and one or two other acts of the kind, proved too much for the patience of the Prussian court; and, in September, 1806, it recklessly drew the sword, amidst the exultation of an army proud of the great traditions of Leuthen and Rosbach. A daring offensive movement was begun; and by the first days of October the Prussian forces had crossed the Elbe, and carelessly advanced, extended along an immense line, from the Lower Saale to the Thuringian Forest. The Grand Army which, since Austerlitz, had remained, for the most part, in Germany, and had been gradually directed on the Maine, was now moved through the Franconian defiles; and, issuing from the valley of the Upper Saale, fell on the rear of its incautious foe, and overwhelmed the Prussians in a great battle at Jena, and another fought on the same day at Auerstadt. This success proved decisive, though Napoleon's manœuvres were hardly as skilful as in previous campaigns; in a few days the whole Prussian army, driven across the Elbe, had either disappeared or become a mass of demoralised captives; Berlin had been opened to the conquerors;

It declares war against France.

Campaign of 1806.

Battles of Jena and Auerstadt, Oct. 14, 1806.

and the French standards had advanced to the Oder, Ruin of the
the military Monarchy of Frederick the Great Prussian
army and
having been crushed in about three weeks. monarchy.

This astonishing triumph, in its rapid sudden- Napoleon
ness surpassing all that he had as yet achieved, im- marches
against the
pelled Napoleon to fresh efforts. Russia had declared Russians.
war before the rout of Jena, and had marched an
army across her frontier; a few thousand defeated
Prussian troops had escaped to the northern verge
of the Monarchy; and, disdaining the perils of a
winter campaign, the victor resolved to press forward, and to bring the war to a speedy conclusion.
His legions were soon upon the Vistula ; and having
crossed that great barrier stream, he endeavoured to
bring his enemy to bay in the vast region of marsh
and forest formed by the Bug, the Narew, the Ukra,
and other rivers of Western Poland. But here his Winter
method of rapid invasions, his troops living on the campaign
in Poland.
territories they entered, received a check from the
forces of Nature, significant of its essential dangers ;
the Grand Army was arrested in its march, and exposed to cruel privations and want in the midst of
barren and pathless swamps ; and, after a series of
fruitless engagements, it fell back from Pultusk to
the Vistula. The French Emperor now put his Campaign
of 1807.
soldiers into winter quarters along the line which
extends from Warsaw to Thorn and the Baltic, and
made preparations to besiege Dantsic ; but he was
not given the repose he expected. The Russian

commander Benningsen, proud of having resisted the conqueror with success, attempted to assail him in his cantonments; and moving his army behind the screen of the lakes which fill the distance from the Narew to the Passarge, fell on the extreme left of the French divisions along the seaboard of Eastern Prussia. Napoleon, however, had antici- pated the stroke; and his antagonist having begun to retreat, he pursued and attacked the Russians at Eylau on February 8, 1807. The battle was terrible and sternly contested; and though the Russians retired from the field, the losses of the French were so heavy that they were not equal to prolong the contest. Napoleon was now in real danger, far away from France, and with the great Powers of Germany conquered, but indignant, oc- cupying his retreat; but he stood firm and applied himself with more than even his wonted energy to restore his forces. Troops were raised in thousands from all parts of the Empire, of which its chief wielded the ample resources with extraordinary ad- ministrative skill; and in a few months the havoc of war was repaired, and the Grand Army in greater strength than before. Hostilities were resumed in June; and Benningsen imprudently advanced to attack an antagonist greatly his superior in force. The Russians were soon repelled from the Passarge; and Benningsen, in an attempt to get back to the frontier, having crossed the Alle with extreme incau-

Indecisive battle of Eylau, Feb. 1807.

Peril of Na- poleon.

Reorgani- sation of the Grand Army.

Decisive victory of the French at Fried- land, June 14, 1807.

tion, Napoleon fell on him with terrible effect, compelled him to fight with his back to the stream, and routed him on the 14th of June, not far from the little town of Friedland. This stroke was decisive; before a week had passed the Grand Army was on the banks of the Niemen; and, with Dantsic, the whole remaining provinces of the Prussian Monarchy passed into the hands of the conquerors. Within less than two years the imperial eagles, which crowned the standards of the French armies, had flown from the British seas, across prostrate Germany, to the distant verge of the Russian empire; war had never been seen in such grandeur before; though Napoleon's movements had not been free from hazards which had attracted the attention of a few thoughtful minds, though unseen by the crowd in the glare of victory.

In this series of triumphs we see the strategy of 1796 repeated, on a larger scale, and with greater results. To seize the decisive points in the theatre of war, to bring a superior force upon them, and to interpose between divided enemies and beat them in detail by rapid manœuvres, were the main objects of Napoleon's movements; and he generally attained them by daring attacks, and by forced marches which placed his soldiers on the most vulnerable parts of the hostile line. In these campaigns, however, he had been greatly seconded by the mistakes of enemies, who had usually contrived to present

Characteristics of these campaigns.

s 2

themselves to his crushing blows; his system, as we have seen, had shown signs of failing when exposed to the strain of natural obstacles; and as the armies he led were infinitely better than those of the Allies in every respect, his exploits were not perhaps so wonderful as those around Mantua and on the Adige.

Changes in the art of war.

Such exhibitions of military force had, however, never been made before; and the antiquated methods of slow advances, of timid movements upon an immense front, and of never passing an untaken fortress, were finally abandoned by European generals. Thus in war, as in many other particulars, the French Revolution wrought changes which had made it a new era in the History of the World; and the strategy of Napoleon, in some of its aspects, was an expression of the increased energy and activity

Meeting of Alexander and Napoleon on the Niemen, June 25, 1807.

generated by that event. The scenes which followed the victory of Friedland rather bore a likeness to a strange romance than to the ordinary arrangements of affairs of State. Unable to resist, the Czar sued for peace; but Napoleon welcomed Alexander as a friend, for he wished to make him subserve his policy; and after interviews between the two potentates, held chiefly in a floating tent on the Niemen, in the presence of the French and Russian armies, peace was made at Tilsit, on the north Prussian

Treaty of Tilsit, July 7 and 9, 1807.

frontier. By this celebrated treaty Prussia was deprived of more than half her former possessions, and became a mere vassal of the French Empire; a kingdom of Westphalia was carved out of her Elbe Provinces and added to the Confederation of the Rhine ; and her conquests in Poland were given to Saxony—she had taken part with France in the late campaigns—under the curious name of the Grand Duchy of Warsaw. At the same time France and Russia united in an alliance of the most intimate kind ; the Czar recognised the French Empire, and pledged himself to uphold its power, and—what was more important—he undertook to offer his mediation to England, and, if she refused it, to go to war with that Power. In return for a co-operation which appeared to set a seal to his domination in the West, Napoleon promised to second the designs of Russian ambition in the North and East ; and he consented to the annexation of Finland, and of the provinces of Turkey north of the Danube, insisting, however, that Constantinople should, in no contingency, become Russian. The conqueror justified the dismemberment of Prussia, and her seeming ruin as a State, by a reference to the proclamation of Brunswick in 1792.

Alliance between France and Russia, and dismemberment of Prussia.

The purpose of Napoleon in making this treaty was to obtain a complete and enduring guarantee for the supremacy of France on the European Continent, to divide Germany more thoroughly than before, and to subject her everywhere to French influence, and, finally, to raise a new foe against England, whose efforts might lead to important

Objects of Napoleon in making the treaty.

results; and it appeared an admirable scheme of state-craft, if such disturbing elements as national passions and the jealousies of rulers had no existence. The dangers, however, that lay hid under the new arrangement of the map of Europe, and in the results of French conquests, were as yet withdrawn from almost every eye; and the power of Napoleon

His power at its height.

was now at its height, though his empire was afterwards somewhat enlarged. At this period that gigantic rule extended undisputed from the pillars of Hercules to the furthest limits of Eastern Germany; if England still stood in arms against it, she was without an avowed ally on the Continent; and, drawing to itself the great Power of the North, it appeared to threaten the civilised world with that universal and settled domination which had not been

Extent of the French Empire.

seen since the fall of Rome. The Sovereign of France from the Scheldt to the Pyrenees, and of Italy from the Alps to the Tiber, Napoleon held under his immediate sway the fairest and most favoured part of the Continent; and yet this was only the seat and centre of that far-spreading and immense authority.

Vassal kingdoms.

One of his brothers, Louis, governed the Batavian Republic, converted into the Kingdom of Holland; another, Joseph, wore the old Crown of Naples; and a third, Jerome, sat on the new throne of Westphalia; and he had reduced Spain to a simple dependency, while, with Austria humbled and Prussia crushed, he was supreme in Germany from the Rhine to the

Vistula, through his confederate, subject, or allied States. This enormous Empire, with its vassal appendages, rested on great and victorious armies in possession of every point of vantage from the Niemen to the Adige and the Garonne, and proved as yet to be irresistible; and as Germany, Holland, Poland, and Italy swelled the forces of France with large contingents, the whole fabric of conquest seemed firmly cemented. Nor was the Empire the mere creation of brute force and the spoil of the sword; its author endeavoured, in some measure, to consolidate it through better and more lasting influences. Napoleon, indeed, suppressed the ideas of 1789 everywhere, but he introduced his Code and large social reforms into most of the vassals or allied States; he completed the work of destroying Feudalism which the Revolution had daringly begun; and he left a permanent mark on the face of Europe, far beyond the limit of Republican France, in innumerable monuments of material splendour. And thus it has happened that much that he founded has survived his fall and his short-lived conquests; the extent of his sway may be still traced by the reach of institutions established by him; and even nations who felt the terrors of his sword and rose justly against his domination, still acknowledge that his rule was not without good, and have a kind of sympathy with the modern Cæsar.

Nor did the Empire at this time appear more

[margin note:] Allied and subject States.

[margin note:] The Empire promoted civilisation in some respects.

Prosperity
of France.
firmly established abroad, than within the limits of
the dominant State which had become mistress of
Continental Europe. The prosperity of the greater
part of France was immense; the finances, fed by
the contributions of war, seemed overflowing and on
the increase; and if sounds of discontent were occasionally heard, they were lost in the universal
acclaim which greeted the author of the national
greatness, and the restorer of social order and welfare. The Jacobin faction had long shrunk out of
sight; the memory of the Revolution and the Reign
of Terror was felt as a foolish or hideous dream; the
public tranquillity was undisturbed; and, in the
splendour and success of the Imperial era, the animosities and divisions of the past disappeared, and
France seemed to form a united People. If, too,
the cost of conquest was great, and exacted a tribute
of French blood, the military power of the Empire
shone with the brightest radiance of martial renown;
Marengo, Austerlitz, Jena, and Friedland could in
part console even thinned households; a career of
glory opened on soldiers which, if brief, was not
seldom brilliant; and the chiefs of the armies,
enriched with the wealth of vanquished Provinces
and subject Kingdoms, and invested with lofty and
sounding titles, forgot the rivalries of an earlier time,
and joined in docile homage to their great master.

Public
works of
Napoleon.
The magnificent public works with which Napoleon
adorned this part of his reign, increased this senti-

ment of national grandeur; it was now that the Madeleine raised its front, and the Column, moulded from captured cannon, which a fresh outburst of Jacobin frenzy overthrew, only a few months ago, in the presence of the mocking enemies of France; and Paris, decked out with triumphal arches, with temples of glory, and with stately streets, put on the aspect of ancient Rome, gathering into her lap the gorgeous spoils of subjugated and dependent races. The government of the Empire had by this time become of a purely monarchic type; it had abolished all republican forms, even to the calendar of 1793, and had made dukes, counts and barons by scores, out of the leading men of the new age; but if it showed the defects of absolute power, it was still essentially the firm, national, and equal despotism of the Consulate, reconciling parties, keeping down anarchy, and as yet acceptable to a people which had forgotten even the thought of liberty. Nor had the great reforms of the Consulate been without ample and beneficent fruit; religious passions had altogether subsided; and the State was administered with an energy, a regularity, and a general equity, which France had never experienced before. *Character of his Government.*

Yet, notwithstanding its apparent strength, this structure of conquest and domination was essentially weak, and liable to decay. The work of the sword, and of new-made power, it was in opposition to the nature of things; it came in conflict *Elements of weakness and decay in the Empire.*

with national traditions, with popular instincts, with moral forces; and it was to illustrate the old fable of the Titans heaping Pelion on Ossa, and being overwhelmed by the bolts of Olympus. The ma-

Indigna-
tion of con-
quered
nations.

terial and even social benefits conferred by the Code, and reform of abuses, could not compensate van- quished but martial races for the misery and dis- grace of subjection; and, apart from the commer- cial oppression of which we shall say a word here- after, the exasperating pressure of French officials, the exactions of the victorious French armies, and the severities of the conscription introduced among them, provoked discontent in the vassal States on which the yoke of the Empire weighed; and made the people of the Confederation of the Rhine, of Germany, and even by degrees of Italy, more or less

Tendency
of Germany
to unite
through
common
suffering.

hostile to the rule of the stranger. The prostration, too, of Austria and Prussia, which had been the result of late events, had a direct tendency to make these Powers forget their old discords in common suffering, and to bring to an end the internal divi- sions through which France had become supreme in

Jealousy of
Russia.

Germany; the recent formation of a Saxon Poland, an evident protest against the Partition, could not fail ultimately to give umbrage to the Czar; and the triumphant policy of Tilsit contained the germs of a Coalition against France more formidable than she

Decline of
the Grand
Army in
strength.

had yet experienced. At the same time, the real strength of the instrument by which Napoleon maintained his power, was being gradually but surely impaired; the imperial armies were more and more filled with raw conscripts and ill-affected allies, as their size increased with the extension of his rule; and the French element in them, on which alone reliance could be placed in possible defeat, was being dissipated, exhausted, and wasted. Add that while it was being thus sapped at the root, the Empire had been continually growing, with a growth too rapid to be sound or lasting, that the ambition of its chief appeared to enlarge as the circle of his conquests expanded, and that the old League of the Continental Powers against the Revolution was being gradually changed into an alliance, unrecognised as yet, but being formed of nations against a military despot; and we shall understand what perils lurked beneath the surface of the imperial sway which overawed Europe.

Nor was the Empire, within France itself, free from elements of instability and decline. The finances, well administered as they were, were so burdened by the charges of war, that they were only sustained by conquest; and, flourishing as their condition seemed, they had been often cruelly strained of late, and were unable to bear the shock of disaster. The seaports were beginning to suffer from the policy adopted to subdue England; and though the Emperor made persistent efforts to prepare for 'an Actium in the British Channel,' they invariably

The re-
sources of
France
unduly
strained.

ended in disgrace and failure. Meanwhile, the continual demands on the youth of the nation for never-ceasing wars, were gradually telling on its military power; Napoleon, after Eylau, had had recourse to the ruinous expedient of taking beforehand the levies which the conscription raised; and though complaints were as yet rare, the anticipation of the resources of France, which filled the armies with feeble boys unequal to the hardships of a rude campaign, had been noticed at home as well as abroad. Nor were the moral ills of this splendid despotism less certain than its bad material results. Too much, indeed, has perhaps been made of the political corruption of the imperial system; for though instruments of new power are peculiarly subject to this baneful influence, it does not appear to have been worse than it had been under the fallen Monarchy; and the France of Napoleon did not parade the shameless dissoluteness and social vices which had characterised part of the republican era. It is also unfair to ascribe to the Empire the want of eminence in art and letters which we see in France during the whole period from 1789 to 1815, for this deficiency was mainly due to the concentration on alien subjects of the energies of the French intellect, even if it be true that the attempts made by the Emperor to remove the dearth were rather calculated to prolong and increase it. Still the inevitable tendency of the Empire, even at the time of

Moral evils of the rule of Napoleon.

its highest glory, was to lessen manliness and self-reliance, to fetter and demoralise the human mind, and to weaken whatever public virtue and mental independence France possessed; and its authority had already begun to disclose some of the harsher features of Cæsarian despotism. This was seen not merely in arbitrary acts, but in suspicious jealousy of any forces or influences not controlled by the State, and in an interference, petty and vexatious alike, even with the arrangements of social life; and the effects were slowly provoking ridicule or discontent, though the murmurs as yet were scarcely heard.

The great and paramount cause, however, of the insecurity of Napoleon's Empire was that its existence hung not only on the life, but on the will of its mighty creator. Without solid foundations abroad, and springing from Revolution at home, it was, in the main, the work of a single man; and it might obviously perish as quickly as it arose by the death of its chief, or through the failure of the gigantic projects which he could design and compass, without the least check on his undivided power. And the Sovereign who wielded this immense authority was a soldier who had hardly known defeat, who stood at the head of gigantic armies, and whose soaring imagination, urged by ambition, was one of the most distinctive of many splendid faculties! And the ruler who had reached

Insecurity of his power which depended mainly on himself

these heights of grandeur had overthrown the old
Order of Europe, and had placed his foot on the
necks of conquered nations ; yet was really sustained
only by the unstable forces of a State recently torn
by Revolution, and by a Nation of which the incon-
stancy had been stimulated of late by every possible
means, which could pass with the rapidity of thought
from enthusiastic devotion to scorn and hatred, and
which, especially at the touch of misfortune, could
suddenly awake from gilded servitude, and with
strange levity repudiate what it had seemed to
revere ! Should that vaulting ambition o'erleap
itself, should disaster overtake the spoiled child of
fortune, should that despotism weigh with too heavy
a burden, in what perils would the Empire be involved, amidst a hostile Europe, and a France linked
to Napoleon chiefly by the frail tie of success ?

CHAPTER XIII.

THE EMPIRE TO 1813.

BEFORE the war which ended with the Peace of Tilsit, the conduct of Napoleon had, in many respects, been that of a great and sagacious ruler. He had, doubtless, even then, given proof of a passion for power, and of grasping ambition, had revealed a purpose to extend still further the domination of France on the Continent, had committed a great political mistake in irritating and defying England, and had justly outraged the opinion of Europe by the execution of the Duke of Enghien; and as the character of man does not really change, though circumstances may largely affect his acts, it is possible to show that, from the beginning of his career, he was one and the same in his essential nature. But History, in judging the leaders of States, can hardly enter into this minute enquiry; and as it contemplates the public life of Napoleon during the first part of his wonderful career, it can excuse much that is more or less blameworthy, and finds more to admire than to condemn. Many of the reforms of the First Consul

Retrospect
of the
policy of
Napoleon.

remain monuments of his great capacity, and his
despotism in France, though heavy from the first,
was also attended with immense benefits; and if he
annexed half-subject Italian provinces, if he increased the influence of France in Germany, if he
precipitated an unfinished quarrel with England;
nay, if he put to death iniquitously a Bourbon
Prince, his policy and conduct in these particulars
may be justified, in a greater or less degree, from
the peculiarities of the time, the condition of Europe,
the traditions of France for long ages, the antecedents of the aggressive Republic, and the violence
and confusion of an era of Revolution. But conquest, and all engrossing power on a scale hitherto
unknown in Europe, had the influence on this extraordinary man which they have exercised on
natures of the same kind; they enlarged the scope
of his daring ambition, and made his sanguine intellect believe that nothing was beyond the reach of
his efforts; and they led him into a series of acts, the
imprudence and inexpediency of which were perceived even by ordinary men, and which hastened,
probably by many years, the ruin of that colossal
dominion, which, however, could not have been last-

It changes
for the
worse after
Tilsit.

ing. We have already seen that the treaty of Tilsit
tended to unite Germany, and even Russia at last, in
hostility against the French Empire; and from this
time forward Napoleon engages in enterprises and in
a course of policy, in which, if we still see his genius

for war, and his skill in administering affairs of State, we more often trace the excesses of mere force, the presumptuous over-confidence of success, and the exaggerated notions of arrogance and pride.

One of the first cares of the French Emperor, after sheathing his victorious sword at Friedland, was to mature a scheme for subduing England, and forcing her to accept a humiliating peace, which had occupied his thoughts for a considerable time, and became known as the Continental system. The Directory had, many years before, attempted to injure British commerce by excluding English and colonial produce from the ports and territories of France and her allies ; and England had retaliated by severe measures which even the occasion could hardly justify. Such acts, however, were trifling compared to the vast plan for ruining England through her trade, which Napoleon conceived in 1806–7, and which forms one of the most striking instances in which despotism has set itself to contend against nature. The Lord and controller of five-sixths of the Continent, he deliberately resolved to close Europe, along its circumference, to access from England; and for this purpose, by two famous ordinances, known as the Berlin and Milan Decrees, he declared that English and colonial merchandise should be confiscated wherever it was found, in the Empire or its allied States, and that even shipping which touched at British harbours should be

The Continental system, 1807-8.

T

included in the general proscription. As France commanded the entire coast from the Baltic to the Mediterranean seas, and Russia fell in with Napoleon's project, the effect of this scheme, if fully realised, would have been to shut out England from all the best markets, to cripple her resources, and to blight her industry; and though it was never even nearly carried out, it certainly did her a great deal of injury. The consequences, however, to the French Empire and its dependencies were to be far more disastrous.[1] The Continental system caused frightful distress in a short time in every maritime town from Riga to Amsterdam and Venice, and blotted, as it were, their prosperity out, by sapping and impeding their trade; and it created general and just discontent in all the allied and vassal States, and even within the limits of France, by depriving millions of the conveniences of life, and by subjecting the mercantile and manufacturing classes to the oppressions and exactions of a host of officials charged to enforce its harsh and unfair restrictions. This system, in fact, was a vexatious tyranny which did palpable and widespread mischief, and brought the sense of wrong home to innumerable hearths; and it quickened the exasperation and animosity of the subjugated but reluctant Continent. This, how-

Its mischievous effects upon the Empire.

---

[1] The ruinous effects of the Continental system in weakening the Empire and impelling Napoleon to fresh conquests are well pointed out in M. Lanfrey's *Histoire de Napoleon I.* vol. iii. chap. 10.

ever, was hardly its principal result; the iniquitous
provisions of his commercial policy being largely
eluded outside France, Napoleon was urged still   It urges
further to stretch the boundaries of his overgrown   Napoleon
Empire, and to proceed to fresh annexations and   conquests.
conquests; and this supposed necessity, in conjunction with the promptings of his ever-growing ambition, contributed greatly to his final overthrow.

The establishment of the Continental system
caused Napoleon at once to turn his eyes to Spain
and the neighbouring kingdom, Portugal, which,
though still independent States in name, had, and
Spain especially, become subject more or less completely to the ruler of France. An event which   Project of
occurred in the autumn of 1807 accelerated a   invading
design already formed, and he resolved to drive the   Spain and
Sovereigns of the Peninsula from their thrones, and   Portugal.
to convert it into a vassal Province. Russia, after
Tilsit, as had been agreed, had offered her mediation
to England, and had declared war when it had been
refused; and as Napoleon, with the consent of the
Czar, had proposed to force Denmark to place her
resources at the disposition of the two potentates,
the English ministry had anticipated the stroke,
and, as the Danes would not give up their fleet, had
caused Copenhagen to be bombarded. On the plea   Napoleon
that this act, the harshness of which was infinitely   dethrones
more apparent than real, permitted him to do what   the House of
he liked in Europe, Napoleon pushed forward an   Braganza,
                                                   Nov. Dec.
                                                   1807.

army on Lisbon, and proclaimed that the House of
Braganza had ceased to reign; and soon afterwards
he gradually introduced considerable forces into
Spain, which took possession of the frontier fortresses
and ultimately advanced beyond Madrid. The dis-

The Royal
family of
Spain
enticed to
Bayonne,
and in-
duced to
abdicate
the throne,
May, 1808.

sensions of the imbecile Spanish Bourbons facilitated
the Emperor's unscrupulous policy; and Charles
IV., the nominal King, having refused to sanction
an abdication in favour of his son extorted from
him, Napoleon induced the whole royal family to
accept him as the arbiter of their rights, and having
brought them together at Bayonne, obtained from
the old King a cession of the Crown, and immediately conferred it on his brother Joseph. This
treacherous deed of violence and wrong, though
accompanied by a new Constitution for Spain, which
put an end to many inveterate abuses, led to unex-

General
rising in
Spain, May,
June, 1808.

pected and momentous consequences. The pride of
the Spaniards was stirred to its depths; the Nation
sprang as a man to arms, to resist the detested yoke
of the stranger; juntas, as they were called, promoted and organised an insurrection in every province; and in an incredibly short time great bands
of levies, far from worthless foes in a land of mountains, had, with what existed of the regular army,
fallen on the invaders wherever they could be found,
and encircled them, as it were, with a consuming
fire. The rising was vigorously supported from
England, and before long its effects were remark-

able. Napoleon had been completely surprised, in his usual scorn of popular feelings; his forces in Spain were widely scattered, and unable to keep the country down; and though his soldiers were easily victorious in one or two engagements in the open field, one of his lieutenants, Dupont, was obliged to surrender with a large detachment in the Sierra Morena, and another was ignominiously driven out of Castile. Meanwhile, a British force under Sir Arthur Wellesley—a name destined to illustrious fame—had defeated the French divisions in Portugal, and had also compelled them to capitulate; and a French squadron in the harbour of Cadiz had been destroyed or forced to strike its colours. Before the autumn of 1808 the imperial eagles had been made to fly in disastrous retreat towards the seat of their power, and not a Frenchman south of the Ebro was seen.

Capitulation of Baylen, July 19, 20, 1808.

First appearance of Sir A. Wellesley on the scene. Convention of Cintra, Aug. 30, 1808.

Great reverses of the French.

These disasters amazed and excited Europe, and filled Napoleon with indignation. His renown, he knew well, was the mainstay of his power; and he poured troops into Spain in thousands to subdue what he called 'a rising of the mob.' His disciplined armies soon scattered in flight the levies that ventured to cross their path; and, having swept through the Somo Sierra pass, he installed his brother in pomp in Madrid. But the national resistance lived on; it broke out in a savage guerilla warfare, in a country made for a movement of the

Napoleon invades Spain and enters Madrid, Dec. 2, 1808.

kind; and Saragossa gave a glorious example of a defence imitated by other cities. Napoleon, however, went on persistently with the work of subjugation; and before long he had crossed the Guadarramas again, in pursuit of a small British army, which from Leon had threatened his line of retreat, but was retiring before his overwhelming forces. His march was interrupted by the news that the attitude of Austria was becoming dangerous; so quitting the Peninsula, he returned to France; and the enemy whom he had hoped to crush not only effected his escape to the sea but inflicted a check on one of his best lieutenants. By this time, however, a fresh contest had begun on another theatre of war. Encouraged by recent events in Spain, and supported by the British exchequer, Austria rose suddenly and declared war; and the Archduke Charles, in April 1809, advanced with a large army from the Inn to the Iser, his object being to surprise the French and their allies, dispersed widely on either bank of the Danube. Napoleon, however, who had arrived from Paris, had just time to anticipate the stroke; and drawing together his scattered divisions with admirable precision, quickness, and art, he turned the Austrian left wing, broke through its centre, and compelled the Archduke to take refuge, completely defeated in a game of manœuvres, not the least wonderful in the Emperor's career, behind the neighbouring hills of Bohemia. The

He leaves the Peninsula at the news that Austria was arming.

Campaign of 1809 in Germany.

Defeat of the Archduke Charles in Bavaria, April 18, 22, 1809.

imperial legions once more poured victoriously down the valley of the Danube; and within a month from the opening of the campaign, Vienna was, for the second time, in their power. Napoleon, however, was not able in 1809, as in 1805, to master the bridges near the capital; and in an attempt to cross to the northern bank of the Danube, and to bring his antagonist there to bay, he met a serious reverse at Aspern, his army being divided on the stream, and a sudden flood having carried away the artificial passages he had made. This disaster, however, was repaired by prodigies of perseverance and skill; and by July 5 the whole Grand Army, with reserves summoned from Italy and the Rhine, had made its way over the firmly-held river, and debouched into the great plain of the Marchfield, from an island in which it had been camped and fortified. The battle which ensued was bloody and terrible; a vigorous effort made by the Archduke against the French left proved nearly successful; but the Austrian centre and right were pierced; and Napoleon, after a desperate struggle, in which nearly 300,000 men fought, stood at last triumphant on the low hills of Wagram. The blow, though not nearly so overwhelming as those of Austerlitz, Jena, and Friedland, was too much for the strength of Austria, and peace, purchased by fresh cessions of territory, was made at Vienna in autumn.

This campaign restored the power of the con-

Reverse of Napoleon on the Danube at Aspern, May 21, 22, 1809.

Batttle of Wagram and victory of the French, July 6, 1809.

Treaty of Vienna, Oct. 14, 1809.

queror; and a subsequent event appeared to increase
it. Napoleon had married several members of his
family into royal Houses—vassal Sovereigns could
not refuse him anything—and after Wagram he
found means to repeat the experiment in his own

Napoleon
divorces
Josephine,
and marries
the Arch-
duchess
Maria
Louisa,
March 11,
1810.

person. His wife Josephine had borne him no child,
and in order to strengthen and prolong his dynasty
he obtained a divorce, and soon afterwards married
a youthful daughter of the Emperor of Austria, the
Archduchess Maria Louisa. This marriage was celebrated with extraordinary pomp, and seemed to set
a seal on his greatness; nor was the advantage for
a time illusory, as Austria, wearied with repeated
defeats, for the present inclined to a French alliance.
England, backed by the insurrection of Spain, now
became once more the only open foe of the master
of more than thirty legions; and had Napoleon, at
this juncture, made the Peninsula feel the whole
power of his arms, he could, humanly speaking, have
subdued the country, and kept it free from English

Successes
of Sir
Arthur
Wellesley
in 1809, in
Portugal
and Spain.

intervention. But the Emperor comparatively neglected Spain, for his armies had routed the Spanish
levies in several engagements in 1809, and, if they
had been foiled by a British force at Oporto, and in
a struggle at Talavera, they had lately compelled
it to retreat into Portugal; and as the conquest
appeared nearly complete they were scattered over
a variety of points, and nowhere collected for a
decisive movement, although in numbers extremely

formidable. This state of things was thoroughly
understood by a commander whose wisdom was to
throw a momentous weight into the scale of fortune,
and who had perceived with profound insight the
weak point in Napoleon's system of war, and the
best method to cope with it. Sir Arthur Wellesley
—become Lord Wellington for his success at Oporto
and Talavera—had seen clearly that the rapid invasions of the French armies without regular supplies
might be encountered by obstacles and want; and
as the forces of Napoleon in the Peninsula appeared
unlikely to draw together, he had satisfied himself
that he could find the means of maintaining himself
against any probable foe, and, in any event, of re-
embarking his troops. For this purpose he caused
a position on the verge of Portugal, between the
Tagus and the sea, to be fortified with extreme care
and secresy; and he gave orders that, should the
French advance, his army should retreat to this
place of vantage, destroying, as it fell back, the
adjacent regions. The consequences of these masterly
arrangements were memorable in the highest degree.
Napoleon, ignorant of what his antagonist had done,
directed Masséna, in the summer of 1810, ' to drive
the English into the sea ; ' but the French army was
far too weak for the purpose; and when, having
been checked at Busaco, it arrived before the im-
penetrable Lines—ever since famous as those of
Torres Vedras—after a march of suffering through

a wasted country, it recoiled amazed from an impassable barrier. After a series of attempts to bring Wellington to bay, Masséna was ultimately compelled to retreat; and he reached the frontier with the mere wreck of an army ruined by disease and privations.

Great results of this campaign, and its influence on Europe.

The issue of this remarkable campaign cause wonder and hope to thrill through Europe. The forces of Napoleon in Spain were immense; and yet the conqueror had been worsted, at the decisive point, by a small army; and a way to encounter him seemed discovered. Soldiers began to study the strategy of Wellington as they had studied that of the French Emperor; and the name of Torres Vedras was in every mouth as that of Rivoli and Arcola had been. Meanwhile, Germany shook fiercely in her bonds; secret societies spread the flame of patriotism, and invoked vengeance on the detested foreigners; and though the courts of Austria and Prussia stood timidly aloof, and the princes of the Confederation of the Rhine continued to lick the hand of their master, the divided members of the great Teutonic family drew towards each other with the feelings of

Agitation in Germany and Holland.

a common nationality and hate of oppression. Disturbances, too, broke out in Holland, half ruined by the Continental system; symptoms of discontent were apparent in Italy; and the tale of Continental troubles was increased by a violent quarrel between Napoleon and the Pope. Nor was France free from

anxious symptoms; Bordeaux, Marseilles, and the
seaport towns·were full of sounds of anger and distress; the steady consumption of war in Spain had
made the conscription extremely unpopular; taxation and bankruptcies had enormously increased;
and notwithstanding a watched Press, and mute or
obsequious Bodies of State, unquiet murmurs began
to be heard and even to threaten a distant tempest.
The real strength of the imperial armies had also
become more and more weakened; the soldiers of   Murmurs
Wagram were very inferior to those who had marched   in France.
from the camp at Boulogne; and the addition of
feeble boys to the ranks, and of auxiliaries listless if
not false, had gone on with accelerated speed. Every
sign, in a word, of coming danger, which could have
been noted in 1807, had grown more visible in 1811;
and the birth of a son at this time to Napoleon,   Birth of a
was not felt to be, in general opinion, as it would   son to Napoleon,
have been a few years before, an assurance of the   March 20,
1811.
stability of his throne. The conqueror, however,
from the height of his splendour could not see the
shadows that were creeping on; and in the face of
the Continent, awed but indignant, he incorporated
Rome, Holland and the Hanse Towns into the territories of the French Empire, in order to complete
his rule in Italy, and to carry out more thoroughly
the Continental system.

These aggrandisements could not fail to arouse
the jealousy of the only State on the Continent

which still preserved a shadow of independence.

Jealousy of
the Czar,
and dis-
putes with
Russia.
The Czar had soon abandoned the alliance of Tilsit ;
he had resented a refusal of the French Emperor to
pledge himself not to restore Poland or to make
additions to the Grand Duchy of Warsaw ; and he
protested against the annexation of the Hanse Towns,
and the occupation of Prussia by French troops
which had been continued ever since Friedland.
Napoleon had retorted by insisting on the necessities
of the Continental system ; and increasing coolness
became open dissension, Alexander having, in self-
defence, relaxed some of its worst restrictions. Napoleon resolved in 1811 to invade Russia the following year; and the preparations he made for the
enterprise surpassed all that he had yet attempted.

Napoleon
prepares to
invade
Russia,
Nov., 1811,
May, 1812.
Austria and Prussia were compelled to promise him
support ; the princes of the Confederation of the
Rhine were ordered to have their contingents ready;
material of war in immense quantities was accumulated in the North German fortresses ; enormous
magazines were formed to afford subsistence to half
a million of men, for the difficulty of following the
usual system of war in Russia had been foreseen ;
and the whole forces of Western Europe were banded
together for an expedition unequalled in its gigantic

Campaign
of 1812.
conception. Slowly and by degrees the prodigious
host, an assemblage of many races and tongues—
Italians, Germans, Dutch, Poles, and even Spaniards
and Portuguese, commingled with the dominant

French—was moved from distant points on the Continent; and by the early spring of 1812 it was aggregated on the plains of Northern Germany. The Emperor, leaving Paris in May, was soon at Dresden, where old Europe, in the persons of humbled and vanquished Kings, bowed in homage before the revolutionary Cæsar ; and on June 24, 450,000 men, with 60,000 cavalry and 1,200 guns, crossed the Niemen from the verge of Prussia, and entered the borders of the Russian Empire. Wilna was attained in a few days ; but the difficulties of the vast enterprise had already made themselves seriously felt ; desertion and disease had set in ; the march of the columns had been delayed by the mass of impediments on their track ; and Napoleon was obliged to make a long halt while the Russian armies, which had advanced to the frontier, escaped from his well-designed manœuvres, and though separated and feebly led retired slowly into the distant interior.

The Grand Army crosses the Niemen, June 24, 1812.

Retreat of the Russians.

The French Emperor had now the means, without incurring any serious risk, of dealing Russia a decisive stroke; he might have avenged a great public crime, and proclaimed the freedom of the Polish Nation. But he characteristically preferred a merely hollow alliance with the Austrian and Prussian Courts to the Polish People ; and though his armies were crowded with Polish soldiers, he intimated to a deputation at Wilna that he could not undo the work of the shameful Partition. He broke up from Wilna

Political mistake of Napoleon in not restoring Poland.

in the middle of July, leaving the greater part of
his impedimenta behind, in order to pursue his
retreating enemies, who, divided into two great
masses under Barclay de Tolly and Bagration, held
an extended line on the verge of Lithuania, from
Drissa on the Dwina, to the heads of the Dnieper.
Napoleon's movements were made slow by bad roads
and the want of supplies, already causing havoc
among his troops; and Barclay, carefully eluding his
foe, succeeded in joining his colleague at Witepsk,
and concentrating the united Russian forces, about
250,000 strong. The Grand Army had, by this
time, lost considerably more than a third of its
numbers, the young soldiers who filled its ranks, and
the auxiliaries, disappearing in thousands; but Napoleon, hoping to outmanœuvre his opponent,
marched upon Smolensko, in order to turn the
positions of the Russians, or to compel them to
fight. Barclay, however, now in supreme command,
and imitating the defensive method of Wellington,
merely checked the Emperor and fell back, destroying the country upon his way, and Napoleon at
Smolensko only found ruins, and a battle continually
eluding his grasp. He resolved still to continue the
pursuit; but it is a mistake to suppose that he took
no precautions; on the contrary he sent large detachments to cover and secure his flanks and rear;
he ordered large reserves to come up from Germany;
and he directed immense magazines to be formed at

Smolensko, Wilna, and other places. Having thus, as he thought, made his advance safe, he set off from Smolensko with about 160,000 men, drawn on through the vast expanses of Russia by the hope of ever-receding victory; but still Barclay stubbornly retired, and the invaders became more and more weakened. At last the indignation of the Russian army at its prolonged retreat led to the removal of its chief, and Kutusoff—the able veteran of 1805— having been appointed to the command by the Czar, was reluctantly forced to offer battle. The encounter took place at Borodino, on the way to Moscow, on September 7. It was murderous beyond all past experience; and though the Russians lost the position, their antagonists could hardly claim a triumph. Kutusoff, however, judiciously fell back, and on September 15, 1812, the Grand Army was master of Moscow, the extreme limit of the march of the Tricolour.

He marches into the interior of Russia.

Battle of Borodino, Sept. 7, 1812.

The Grand Army enters Moscow, Sept. 15, 1812.

The daring advance into the heart of Russia, enormous as the losses of the French had been, now seemed justified by the event; and Napoleon expected to dictate peace. A tremendous catastrophe was, however, to show what patriotism and hatred could plan and accomplish. The governor of Moscow set fire to the city in order to cut off its resources from the French, and as it was chiefly constructed of wood, it was soon a desert of devouring flame. Napoleon, however, still lingered on the

The Russian governor of Moscow sets fire to the city.

Napoleon
delays in
the hope of
peace.
spot, convinced that the Czar would yet treat;
though Kutusoff, meanwhile, with prudence and
skill, had drawn together his shattered forces, and
was already menacing the Emperor's retreat. The

Beginning
of the re-
treat from
Moscow,
Oct. 19,
1812.
hope of negotiation having proved fruitless, the
Grand Army at last left the ruins of Moscow on
October 19, the intention of Napoleon being to
march southward, and to attain Lithuania through a
country in which his soldiers could find the means
of subsistence. The movements, however, of the
French were sluggish, for they had loaded themselves with the spoils of Moscow; and after an indecisive action at Malo Jaroslavetz, the Emperor
abandoned his previous design, and retreated by the

Horrors of
the retreat.
way he had before advanced. The sufferings of the
French in this wasted region became gradually more
and more intense; famine, aided by cold, destroyed
thousands; Kutusoff hung on the flanks of the
perishing host, annoying it with bristling swarms of
Cossacks; and the Grand Army which, before leaving Moscow, was still more than 100,000 strong,
dwindled into a mass of 40,000 fugitives before it
reached the remains of Smolensko. News of fresh
misfortunes were here received; the magazines had
been hardly formed; two Russian armies, bearing
before them the detachments he had left to protect

Imminent
peril of Na-
poleon and
the remains
of the army.
his flanks, were gathering to bar the Emperor's retreat, and the only chance of safety was to press
onward, and endeavour to open a way to Wilna.

The wreck of the Grand Army, before long joined by the divisions which had tried to cover its wings, toiled feebly along the Lithuanian wastes, pursued, as hitherto, by its relentless foes ; and, after increasing losses and horrors, it found itself on the Beresina, assailed and almost surrounded by hostile forces. It ought to have been destroyed to the last man ; but its remains were saved by the skill of its chief, and the terror his name still spread around ; and, strewing its path with the dying and the dead, it gradually approached the still distant frontier. At Smorgoni Napoleon gave the command to his brother-in-law, Murat, the new King of Naples, and hastened off to France to raise fresh levies—a step which has been very differently judged—and after he had gone, the dissolution of the ruined array went on more rapidly. Considerable reserves, which had come up, were involved in the fate of the survivors of Moscow ; and, after plundering the magazines at Wilna, the thinned remnants of the once mighty host repassed the Niemen in little knots and bands, of which some were rallied behind the Vistula. More than 550,000 men, including reserves, had entered Russia, and it is doubtful if 50,000 of these were ever seen again with the eagles.

Passage of the Beresina, Nov 25-28, 1812.

Napoleon leaves the army for France, Dec. 5, 1812.

Destruction of the Grand Army.

The causes of this tremendous ruin, the near prelude to Napoleon's fall, may be indicated in a few words. Something may be ascribed to the bad composition in every respect of the Grand Army,

Reflections on this catastrophe.

U

and something to the effects of the cold; and the conduct of Barclay, after Smolensko, and of Kutusoff, during the retreat, was able. The constancy, too, of the Russians was great; and the burning of Moscow certainly had immense, and possibly decisive, results in depriving the invaders of winter quarters. Napoleon may also have shown a want of his usual energy at Maroslavetz, and perhaps on one or two other occasions; he probably ought not to have left his army; and his manœuvres to overwhelm his enemies failed, though marked by his accustomed skill. The paramount cause of the disaster, however, was that Napoleon's system of daring invasion was adopted on an extravagant scale, and was encountered, after some faulty operations, by the Russian commanders in the fitting way; the Grand Army perished from want, led on hundreds of miles in a barren country; and, curiously enough, the very means which Napoleon employed, at the outset of the campaign, to ensure its support only led to mischief, for its impedimenta paralysed brilliant manœuvres, which otherwise might have brought the war to an end. The example of Wellington at Torres Vedras contributed thus to this mighty overthrow; but Napoleon was anything but the madman which he has been called by superficial critics; and it is at least doubtful whether he would not have triumphed, had not Moscow been suddenly destroyed —a contingency on which no leader could reckon.

Causes of the ruin of the French.

He was a great commander in 1812, as he was throughout his military career, though his over-confidence was more apparent then than it had been on previous occasions; and, apart from the enterprise itself, undertaken in the pride of ambitious power, the chief mistake he probably made in the campaign was that of a politician, not of a chief of armies— the not disarming the Czar on the frontier by liberating the Polish race from its chains.

CHAPTER XIV.

FALL OF NAPOLEON.

Return of
Napoleon
to Paris.

AFTER leaving the remains of the Grand Army—he had hoped that it would rally at Wilna—Napoleon returned to Paris in disguise through the frozen plains of Poland and Germany. The reception he met was very different from that which had greeted the warrior of 1799; and though the official *noblesse* of the Empire concealed their alarm by increased servility, France maintained that silence which is often ominous. An incident during his absence had revealed how really precarious was his Revolutionary

Conspiracy
of Malet.

throne; an obscure republican of the name of Malet had deceived persons in high places by the news of the Emperor's death; and Napoleon heard with amazement and anger that even the Bodies of the State had never thought of his infant son as his possible successor. To strengthen his dynasty, on paper at least, he declared the Empress Regent in the event of his death; but graver matters soon engrossed his thoughts. The Prussian Contingent in the Grand Army, having advanced only a short way into Courland, had made good its retreat comparatively intact; and, on being apprised of the issue of the campaign, its commander, York, openly revolted from the French, and went over with his men to the Russian camp., This defection proved the shock that lets the avalanche loose, and sets it in motion to change the landscape. Northern Germany rose as a man to arms; the Prussian army —it had been organised after Jena on that peculiar system of which we have seen the astonishing results; and it was even now capable of large expansion, though, in deference to Napoleon's jealous will, reduced to a small standing force—compelled its ruler to declare war against France; insurrections broke out in the Hanse Towns; and the heave of a great national stirring was seen in Saxony and the States of the Confederation of the Rhine, and even in Austria under absolute rule. The Czar, who had followed the march of Kutusoff, encouraged this universal movement; and in the first months of 1813 the Russian and Prussian armies, seconded by a great wave of popular war, were sweeping over the North German plains, and effacing the signs of French domination. Murat fled with the ranks of the Grand Army, giving up the command to Eugène Beauharnais, the viceroy of the Italian Kingdom; and that chief, conducting the retreat with skill, and undismayed by the flood of enemies, made good with difficulty his way to the Elbe, though obliged

Defection of York, Dec. 30, 1812.

Rising of Germany, January-March, 1813.

Retreat of the French, February-March, 1813. Energy of Napoleon.

to abandon the French garrisons in the fortresses on the Oder and Vistula.

Napoleon heard this intelligence with scorn and wrath, and addressed himself to pluck safety from danger. He treated the rising of Germany with contempt, as he had treated the rising of Spain, warned his crowned vassals to be on their guard against what he called ' a Jacobin movement,' and to have their contingents ready by the spring; and wrote to his father-in-law, the Emperor of Austria, already hesitating in the French alliance, that he reckoned with confidence on Austrian aid. His real power, however, was of course in France; and he strained his great faculties to the utmost to repair the disasters of 1812, and to make preparations for a fresh struggle. The resources of France were still great; she still lavished them to maintain her power, though cruelly stricken and discontented; and by summoning old soldiers to the eagles, by making regular troops of the National Guards, and by anticipating the conscription of the succeeding year, Napoleon set on foot, in a few weeks, the enormous mass of half a million of men, and even gave it military organisation and form. These levies, however, though still bearing the honoured name of the Grand Army, had little in common with the bands of Austerlitz; every arm, especially cavalry, was weak; and though under a great commander they were to show that they could gain battles, they

His immense preparations to repair his fortunes.

Bad condition of the French levies.

formed a very imperfect instrument of war. The Emperor took the field in the last days of April, and in a short time the survivors of the awful retreat, drawing from the Elbe to the Elster and the Saale, had joined in Saxony the new legions which had, as it were, sprung from the earth at the bidding of their renowned master. By this time the Russian and Prussian armies had crossed the Upper Elbe and exposed themselves to the Emperor's blows, in the hope of gaining the support of the vassal South German States; and they fell on the French as they were advancing on Leipsic, through the broad plains of Lützen. The encounter was stern, but skill prevailed; and though the success of the French was really trifling, the trained soldiers of the Allies were forced to retire before troops composed largely of young conscripts. The star of Napoleon seemed now to emerge in splendour again from passing clouds; the subject Kings of Bavaria and Saxony made haste to put their contingents in his hand; he entered Dresden in a few days in triumph; and, as the Russians and Prussians continued to fall back, he pursued them to the verge of Silesia, and defeated them in a great battle at Baützen, one of the most remarkable of his wonderful exploits. He had now approached the Oder and Vistula, and had he prolonged this victorious march he would certainly have set his garrisons free, and perhaps have crushed for a time the rising of Germany. He thought, however,

Campaign of 1813.

Battle of Lützen, May 2, 1813.

Battle of Baützen, May 20-21, 1813.

Success of Napoleon.

that a delay of a few weeks would greatly improve
his unformed armies; and confident that a decisive
triumph would lay his enemies prostrate at his feet,

Armistice
of Pleis-
twitz, June
4, 1813.

he consented, Austria having intervened, to an
armistice which, as events turned out, was a capital
and striking political mistake.

The negotiations that followed form a signal
proof how ambition and pride may blind genius.
Austria evidently at this moment held the balance
between the belligerent Powers in Germany; and
though the Austrian Germans wished for war with
France, the Cabinet of Vienna, after Lützen and

Austria
proposes
terms to
Napoleon
which he
unwisely
rejects.

Baützen, thought peace with Napoleon an essential
object, and proposed terms which would have left
him master of France, Italy, Holland and Belgium,
providing only for the independence of Germany,
and the suppression of the Confederation of the
Rhine. Napoleon, however, had no desire for
peace; he had established his armies along the
Elbe in positions where he hoped to renew the
glories of 1796 on a grander scale, and to defy even
all Europe in arms, when his military strength
should have been more developed; and he refused
to listen to the proposals of Austria, apparently indifferent to what her forces might be, to the disaffection of the Germans in his ranks, which had
become manifest for some time, and to the national
and angry rising already threatening on every side.
· In this state of things the Austrian Government,

forgetting the old dislike of Prussia, and the recent
ties that bound it to France, and yielding to imperious public opinion, inclined gradually towards the
Allies ; but no engagements were definitively formed,
until events on a distant theatre of war determined
a hitherto halting purpose. After Torres Vedras,
Wellington had fought with varying success in 1811;
but in the following year—his own forces having
been increased by Portuguese levies made good
soldiers by his skilful hand, and drafts from Russia
having weakened the French—he invaded Spain,
winning a great battle at Salamanca upon the
Tormes; and though ultimately obliged to retreat,
he liberated a considerable part of the Peninsula. In
1813 he dealt the decisive stroke; advancing from
Portugal with an army superior in numbers for the
first time to its foes, and aided by masses of Spanish
levies, he routed the French with great loss at
Vittoria ; and in a short time he had reached the
Pyrenees, and stood on the verge of that mighty
Empire which had seemed invulnerable a few months
before. This splendid success decided Austria ; she
threw in her lot with the Allies ; and the most formidable Coalition she had ever yet encountered,
encircled France, already worn out and exhausted.

The successes of Wellington in Spain decide Austria to join the Coalition.

Battle of Vittoria, June 21, 1813.

The French driven from Spain.

Napoleon, it is unnecessary to say, made a fatal
mistake in rejecting these terms; even if he believed that Austria was false, his conduct was arrogant and over-confident. Hostilities began on

Europe in arms against Napoleon.

His views
and objects
in the
contest.

August 10, on an immense circle from the Oder to the Elbe, and from the Bohemian range to the Baltic, the centre of operations being the plains that form Saxony and the south of Prussia. Napoleon, as we have seen, had occupied the Elbe, and held its passages in great strength, throwing out secondary forces as far as the Elbe and Oder on either side; and from this position he hoped to defeat his enemies, and repeat the dazzling strokes by which he had ruined Wurmser and Alvinzi in detail. The conditions, however, of the contest had changed; it was more difficult to reach divided enemies in the broad space between the Oder and the Elbe, than in the narrow area between the Tyrol and the Adige; the allied commanders had learned the Emperor's game; and, above all, the levies of the French were very inferior to the allied armies, composed largely of seasoned troops fired by a sentiment of national

Plan of the
Allies.

hatred. The general plan of the Allies was to avoid Napoleon when he attacked in person, but to fall on his most distant lieutenants, and gradually to converge on their dreaded adversary when his strength had been thoroughly impaired; and as even in numbers they were greatly superior, about 550,000 to 360,000 men, they justly calculated on success at last. Their first movements, however, were ill-designed, and gave Napoleon a brilliant victory which in previous campaigns might have proved decisive. In the absence of the Emperor, who had set off against the Prussians in Upper Silesia, the Austrians and Russians under Schwartzenberg moved through the Bohemian hills on Dresden; but their operations were uncertain and slow; their great antagonist had time to return; and they were completely defeated in a pitched battle, in which Moreau, who had joined their ranks, from animosity to the ruler of France, met a death unworthy of a French commander. *Battle of Dresden, Aug. 27, 1813.*

Napoleon thought that he had now the Coalition in his power, but he was to be taught by a striking example how firm was the constancy of his present enemies. He despatched a force through the Bohemian passes to intercept the retreat of the Allies; and, in the days of Marengo and Rivoli, the manœuvre would probably have been successful. Either through his own over-confidence, however, or from errors in which his lieutenants fell, the detachment was too weak to make victory certain; and it was crushed at Culm by an attack of the Allies, who, instead of surrendering, as had been expected, assailed the French with determined energy. This victory redressed the balance of fortune, and events followed which turned the scale. Adhering to their scheme, the Allies fell on the distant lieutenants of the Emperor; one was defeated on the Katzbach in Silesia, another at Grossbeeren in Prussia, and a third with crushing effect at Dennewitz; and, instead of being rent asunder by his blows, the firm arrays of the allied armies drew in gradually their immense *Battle of Culm, Aug. 30, 1813.* *Battles on the Katzbach, at Grossbeeren, and* circle, and gathered upon their hemmed-in foes. Meanwhile the Grand Army was fearfully diminished by losses in the field, disease, and want; the Con-

at Denne-
witz, Aug.
23 to Sept.
5, 1813.

federate Princes of the Rhine grew doubtful, and gradually assumed a menacing attitude; the auxiliaries deserted from the French in thousands, and vast masses of insurrectionary levies hung on the skirts of the dwindling host, keeping up a ruinous and unceasing warfare. The time had come at last for more daring movements; and Blücher, the vigorous chief of the Prussians, with Bernadotte— once a Marshal of France, but now transformed into a Swedish Prince—crossed the Elbe in the last days of September; while Schwartzenberg issued again from Bohemia, the object of the Allies being to meet at Leipsic and overwhelm their adversary. Napoleon, had his movements been free, might perhaps even yet have baffled his foes; but he could not trust his vassals in his rear; and he was slowly but surely forced upon Leipsic, and compelled to

Great
battles of
Leipsic,
Oct. 16 and
18, 1813.

fight at great disadvantage. The first encounter took place on October 16; and though the Allies were at least 230,000 strong and the French not more than 150,000, the terror inspired by the Emperor was such that the battle had no decisive result. By the 18th, however, great reinforcements had poured in to support the Allies; the Saxon contingent abandoned the French on the field of battle, and fiercely attacked them; and, after a desperate conflict, the Grand Army, which fought magnificently when brought to bay, was gradually compelled to leave Leipsic. The destruction of the single bridge on the Elster, on the line of retreat, caused frightful confusion; a large part of the French army was cut off; and the victor of many fields was driven to the Rhine, leaving his garrisons on the Oder and Vistula to their fate. A gleam of success shone feebly on the retiring host; Bavaria had joined the Coalition, and Napoleon crushed a Bavarian force that had placed itself recklessly on his path; but in the first days of November the allied standards, borne by the power of embattled Europe, lowered on the imperilled Empire from across the Rhine. *The French driven to the Rhine.*

Such had been the results of the campaign in Saxony; and though the defections of the German troops, which contributed largely to the final issue, might silence those who have been lately holding up French military honour to the scorn of Europe, Germany had been set free from foreign invasion, and her people had shown heroic patriotism. In other parts of the theatre of war, fortune had also turned against the French Emperor. Austria had invaded Italy from the north; Eugène Beauharnais had been beaten on the Adige; and Wellington, after a vigorous conflict with Soult, one of the ablest of the Imperial lieutenants, had descended on France from the Pyrenean frontier. Thus war gathered *Defeats of the French in Italy. Wellington invades France.*

from all sides on the Empire, and the internal condition of that huge structure showed ominous signs
of collapse and ruin. The Princes of the Confederation of the Rhine had before this abandoned their
master; the Kingdom of Westphalia had already
fallen; and Holland, and even Belgium, wasted by

Revolt of the allied and subject states. the conscription of the Continental system, had
either risen or threatened revolt; while far to the
South, Murat—'a paladin in the field, and a fool in
the closet,' in Napoleon's phrase—was trafficking
with Austria, to save Naples. In France, too, the
late mistress of Europe, the state of affairs was
extremely alarming, and every thing portended ap-

Desperate condition of the Empire. proaching disaster. 1813, following 1812, and the
devouring years of the Spanish war, had consumed.
the military strength of the Nation; and only the
shadows remained of the proud legions which had
once trampled on prostrate Europe. Even the
material of war was wanting in old France; it had
been dissipated on a hundred fields, or transported
to the Adige and the Elbe; and the finances, once
upheld by the spoils of conquest, had suddenly
failed, and were wholly exhausted. Nor was the
temper of the Nation such as could endure invasion
or continued defeat; its ardour of 1793 had died
out; long wars and despotism had impaired its
energy; and it was rather overwhelmed by the sense
of misfortune, than resolved bravely to meet and
subdue it. Though, too, the numerous classes and

interests enriched by the Empire still clung to it, they secretly felt the general discontent; and the very servility of the instruments of power added to the dangers arising from the instability of a Revolutionary State, and the mobility of the national character.

Napoleon did not yield to despair, though ruin seemed on all sides imminent. He might at this juncture have obtained peace by ceding part of the frontier of the Rhine; but he thought of little but a death-struggle. He gave orders to summon to the field all Frenchmen who had served in the army, though he characteristically refused to appeal to the nation; and, calculating that the Allies would not move till spring, he prepared to contend for the greater part of the Empire. The remains of his forces were distributed along the immense front, from the Scheldt to the Adige, as he believed that he would have time to reinforce them; and though he finally abandoned Spain, he resolved to strike for the whole Rhine and Italy. Had he been permitted to mature his plans, it is difficult to say what the result might have been; but the Coalition had been taught not to repeat the errors of 1793; and the allied chiefs were in a very different mood from the Yorks and Brunswicks of a former day. Towards the close of December 1813 they set in motion their immense hosts; and Blücher and Schwartzenberg crosssed the Rhine in two masses from Coblentz to

Napoleon thinks only of a death-struggle.

His preparations.

The Allies invade France, Dec. 20–26.

Bäsle, while to the North Bernadotte invaded
Belgium, and Wellington, to the South, advanced to the Adour. This sudden and overwhelming invasion completely disconcerted Napoleon's
projects, and for several weeks met no resistance on
the theatre where it was most formidable. Driving
before them some feeble French detachments, and
masking, as it is called, the fortresses on their way,
Blücher and Schwartzenberg soon passed the Vosges;
and by the middle of January 1814, their converging armies reached the edge of the vast plain which,
watered by numerous streams, extends through
Champagne to the capital of France, from the
western hills of Lorraine and Franche Comté.

*The mili-
tary situ-
ation of
Napoleon
seems hope-
less.*

The military situation of the French Emperor
at this juncture appeared hopeless. He had raised
only a small part of the levies he had intended to
collect; and he had probably not 250,000 men, including the remains of his Spanish armies, to oppose to the hosts of the Coalition, which numbered
fully 500,000, supported by enormous reserves.
His troops, too, were in part worn out and demoralised, and his lieutenants had lost their wonted
confidence; while the allied commanders were
flushed with success, and their armies burned with

*Prostration
of France.*

fierce national passions. France also seemed without hope and prostrate; and even the obsequious
Bodies of State, and the new *noblesse* of the Revolution, had began at last to show dangerous symp-

toms of open insubordination and anger. Napoleon, Campaign of 1814. however, did not despair, and prepared to encounter Blücher and Schwartzenberg, though these leaders had more than 200,000 men, within easy reach of each other in Champagne, and he had hardly more than 70,000 in hand. His first operations were un- Battles of Brienne fortunate ; in a daring attempt to separate the Allies, and La Rothière, he fought an indecisive battle at Brienne, and was Jan. 29, beaten with heavy loss at La Rothière ; and had his Feb. 1, 1814. antagonists followed up their success, or even acted with ordinary skill, they could have made the issue of the campaign certain. But Blücher and Schwart- zenberg had advanced on divergent lines, and were alienated by mutual dislike and jealousy ; and, ac- cordingly, at this critical moment they divided in- stead of uniting their forces, and began to march on Paris by distant roads, one along the Marne, the other along the Seine. The opportunity was not lost by the Napoleon interposes great warrior who stood in their path, and whose between the Allies. powers were never, perhaps, more evident than when in a position of this kind. Availing himself with consummate art of the obstacles formed by the double rivers, and leaving a detachment to hold Schwartzenberg in check, Napoleon, in the first days of February, marched against Blücher, who had spread his forces along the Marne with careless con- fidence ; and the result was worthy of the General of 1796. Breaking in on the side of the Prussian army, Napoleon met its separate divisions, and mul-

Battles of
Champ-
aubert,
Mont-
mirail,
Vau-
champs and
Nangis,
Feb. 10–18,
1814.
tiplying his swift and terrible strokes, routed them
one after the other in detail, at Champaubert, Montmirail, and Vauchamps; and in less than a week
the discomfited chief was driven, completely beaten,
on Châlons, with forces reduced to half their numbers. The Emperor now turned against his second
enemy, descending from the Marne to the Seine;
and in a short time the army of Schwartzenberg,
which had also pressed forward with little caution,
was pierced through and compelled to retreat, after
a double defeat at Montereau and Nangis. The
losses of the Allies had been so great, that Schwartzenberg actually sought an armistice; and at the
close of February the invading host had fallen back
to the positions in Champagne, from which it had
moved a month before.

Astonish-
ing success
of Napo-
leon.
These operations rank justly among the finest
specimens of Napoleon's skill, though made possible
only by the errors of his foes. Negotiations were
now set on foot, and had he abandoned Belgium and
Italy, he might have preserved part of the revolutionary conquests; but he refused, either from in-
domitable pride, or confidence in his late extraordinary success. The Coalition resolved to continue the war; and events on other parts of the
Success of
the Allies
in other
parts of the
theatre.
theatre contributed largely to confirm their purpose.
The arms of Wellington progressed in Gascony;
Eugène Beauharnais was being driven from Italy;
and Murat, with the disloyalty of a revolutionary age,

was actually preparing to march from Naples, and make common cause with the allied armies. It was, therefore, evident that the British commander would occupy a large part of the Imperial forces—the Army of Soult at this juncture was in fact superior to that of his master—and that a fresh attack would be made from the east; and it was thought impossible but that the allied armies would at last crush their still dreaded antagonist. Hostilities were resumed in the beginning of March; and, in order to make success certain, Bernadotte was directed to advance to the Meuse, Schwartzenberg refusing otherwise to move; though united to Blücher, he was still immensely superior to the French Emperor in force. Napoleon proceeded to renew against Blücher his late manœuvres; and he had nearly caught his stubborn foe, who, though daring to a fault, was wanting in skill, when Blücher was saved by the surrender of Soissons, and having joined the vanguard of Bernadotte, was able to offer battle in preponderating strength. Napoleon was compelled to recross the Asine, after a bloody and disastrous action at Laon; and having thus failed to defeat Blücher, he thought himself unequal to assail Schwartzenberg; and threatened with destruction by their uniting armies, he formed a resolution which, though fatal in the event, was worthy of his art as a military scheme; and, in other times, might have proved successful. Considerable forces were

Fresh forces raised against Napoleon.

Battle of Laon, March 9–10, 1814.

x 2

locked up in the fortresses on the Meuse and the Moselle—those on the Oder and Vistula ,had been

Napoleon
falls back
on Lor-
raine, to
rally his
garrisons,
and strike
the rear of
th Allies.

lost—and the Emperor determined to fall back on Lorraine, to add these garrisons to his army in the field, and then, descending on the rear of his foes, with a force stronger than he had yet possessed, to oblige them to fight in a position in which a single defeat might prove as ruinous as that of Melas had been at Marengo. He broke up from the Aube towards the end of March, after a short conflict with the enemy on his way; and, concealing the movement by a screen of horse, his columns sought the roads to the Moselle.

This march of Napoleon would have certainly made the Allies pause, on ordinary occasions, and might have exposed them to his strokes ; but though Blücher and Schwartzenberg had suffered heavily, the Coalition held firmly together, and national

The Allies
march on
Paris,
March 25,
1814.

passions inspired its armies. At a council of war held on March 24th, it was resolved to disregard the Emperor's movement, and to make a great effort to bring the war to a close, by marching directly on the capital. The condition of France, and of Paris itself, concurred to favour this bold design. The Nation, utterly exhausted by war, had become wearied of the Imperial rule ; the distress of most of the great towns had caused the royalist and republican parties, long silent, to raise again their heads; and in the capital, the centre of thought and opinion, Napoleon's tottering throne was mined by intrigue. A sentiment had spread that could peace be obtained, and the interests of the Revolution be saved, the Emperor ought to be made a sacrifice; and it had made way among the aristocracy of wealth, which had worshipped Napoleon in the day of success, among the Bodies of State, which, in this manner, avenged themselves for the slights of power, and among the masses of a thoughtless populace demoralised by the events of the last twenty years. Thus everything led the Allies to believe that the fate of Paris would prove decisive; and their great armies were set in motion, converging upon the defenceless capital, which for so long a time had been the ardent focus of trouble, disturbance, glory, and empire. Driving before them a few weak bodies of troops which attempted in vain to retard their advance, they had soon reached the hills overlooking Paris; and after a brief but sharp struggle the city surrendered on March 30. The hopes of the Allies were soon verified: on an assurance that the rights which had grown up since the Revolution would be guaranteed, the once humble and flattering Senate declared the Crown of Napoleon forfeited; the example was followed by the different Bodies which represented the Nation or the State; and, as in the presence of the hosts of Europe, no other choice could have been accepted, the Bourbon Monarchy was re-established in the

State of opinion in the capital.

Capitulation of Paris, March 30, 1814.

Napoleon dethroned, the Bourbons restored.

person of the Count of Provençe, the second brother of Louis XVI. Some interested demonstrations of joy were made; but though the Nation, on the whole, acquiesced, and changed the Empire with the same suddenness with which it had changed the extinct Republic, it felt intensely the humiliation of defeat, and received the Bourbons without sympathy; nor did thousands forget the name of Napoleon even when, under the stress of crushing disaster, it was widely denounced as the symbol of ruin.

Napoleon hastily retraces his steps.

While these memorable events were occurring, the Emperor had pursued his march eastwards; but on the news of the allied movement, he retraced hastily his steps through Champagne. He arrived at Fontainebleau, with about 70,000 men, as the capitulation was being signed; and for a moment he formed the desperate design of falling on the Allies, who had divided their forces negligently upon the Seine, in the confidence of assured success. His lieutenants, however, protested against an attempt which might have destroyed Paris, even though, as he insisted to the last, it was promising from a military point of view; and one of them, Marmont, having, without their knowledge, placed his divisions in the power of the Allies, the conqueror's sword fell broken from his hand, and he was left defenceless in

He abdicates, April 4, 1814.

the midst of his enemies. In a few days he abdicated the throne; and the fallen Lord of five-sixths of Europe, deserted by those whom he had raised to greatness, though his soldiery clung with devotion to their chief, was left to muse, unheeded and alone, on the instability of human things, and the punishment of unbridled pride and ambition. The small island of Elba had been given him in exchange for the Empire he had lost; and, after a touching farewell to the veterans of his Guard—the Tenth Legion of the modern Cæsar—he set off for his insignificant realm, the populace of the maritime towns having more than once beset him, on his way, with execrations which made him feel the misery caused by the Continental system. Thus fell from the loftiest height of grandeur attained by man in the modern world, that mighty product of the French Revolution—the Lucifer, as he has been called with some truth, of the gigantic strife of the first part of the century. Those who regard Napoleon as a mere tyrant, destructive, cruel, inhuman, selfish, see only a very small part of his character, and pervert it by this imperfect estimate. Many as were his faults and, we may say, his crimes, this wonderful being conferred benefits on France which she has not forgotten; and if his despotism was an evil from the first, and contained the germs of future disaster, and if his ambition was always perilous, his government was able and moderate for a time, and even his bloodstained career of conquest was not without good results in Europe. His fall is the old tale of the terrible effects on the conduct of man of un-

Character of Napoleon.

bounded power; and the potentate who, after the
Treaty of Tilsit, set himself to oppose the laws of
nature, invaded Spain with perfidious insolence,
plunged into the frozen deserts of Russia with
Europe conspiring on his homeward path, and preferred to challenge the world to arms to the surrender of a worthless ascendency, seems a different
person from the Bonaparte of Luneville and the
author of the Concordat and the Code. For the
rest, if Napoleon had few scruples, and was pitiless
in carrying out his aims, this may be accounted for,
in some measure, by the moral confusion of the
France of his time; and if he made self the centre
of his hopes, he associated self with national greatness. As a General he created modern war; and
though his passionate and daring imagination made
him over-confident as a military chief, and his
strategy of invasion was not always safe, he stands
preeminent as a leader of armies, was a master of
his art in all its departments, and was wholly unrivalled in those great combinations which form the
highest problems of military science. His greatest
fault as a politician was the contempt of national
feelings and instincts, which led him into innumerable mistakes; nor did he, perhaps, give proof of
the gifts which distinguish statesmen of the first
order; but he had good reason to despise and distrust the popular movements of the France of his
youth; and he possessed in the very highest degree

the faculty of administration, and even of government. Let it be added, too, that perhaps his despotism was inevitable in the existing condition of France, that for years it was the glory of Frenchmen, and that, to this day, it has been, in part, justified by the noble institutions and great measures, with which History will always connect it. The offspring of the Revolution and yet its controller, Napoleon stands on the tracts of the Past, the most prominent figure of a wonderful age; and the shadow of the great name along the path of Time seems to blight the pretensions of rulers alien to his own race in the land he swayed.

In the readiness of France to throw off Napoleon we see a fresh proof of the national character; and the manner in which French officials of State and dignitaries of every kind abandoned the master to whom they owed everything, stands in marked contrast to the stedfast loyalty of Austrian and Prussian nobles to their Kings after such calamities as Jena and Austerlitz, and to the constancy of the Allies in 1813–14. Before, however, we censure Frenchmen generally, all the circumstances must be taken into account, and condemnation must be largely qualified. After making efforts such as never, perhaps, have been made by a European State, France was utterly broken down when the invasion came; and in this condition of affairs we can hardly feel surprise that she deserted a Sovereign who, at the moment,

Reflections on his fall.

appeared the existing cause of her sufferings and whose chief title to her obedience was success. As for the conduct of the marshals and ministers who forsook Napoleon in the hour of misfortune, it was such as has more than once been seen in the case of a mere *noblesse* of functionaries, the new-made instruments of new-made power, and without the traditions, and the sense of honour, that distinguish an aristocracy worthy of the name. Apart, however, from the national temperament, the inevitable result of the Revolution was to weaken in France every tie that binds the State and even society together; and, accordingly, when it was put to the proof, the authority of Napoleon suddenly collapsed, and could not bear the strain of disaster, the truest test of institutions and men. Still we must not imagine that all classes were indifferent to the fall of the Empire; the remains of the Army mourned their chief, and his name retained its spell in parts of the country. Nor can we ascribe to the Revolution alone the precarious nature of his unstable rule, for the Monarchy of the Bourbons was overthrown with greater facility than the Empire, and left, perhaps, fewer adherents behind. In fact, the corruption of the old order of things had blighted loyalty and faith in France before the events of 1789; and we must not ascribe the whole difficulty of establishing power in that country to the period of disorder that followed, though this has certainly been a principal cause.

We must add, too, that it was not only those of new origin, and recent dignity, who betrayed Napoleon or fell away from him; his imperial consort shook him off as lightly as she would have shaken off a disagreeable dream; his discarded plebeian wife died of a heart broken ' at the ruin of her Cid.'

CHAPTER XV.

THE HUNDRED DAYS AND WATERLOO.

Peace of
Paris, May
30, 1814.

FRANCE, after the capitulation of Paris, was at the mercy of the victorious Coalition. Owing, however, to the interposition of England, the conditions of peace were less onerous than the vanquished Nation might have expected; though stripped of all her revolutionary conquests, she was left with her ancient boundaries intact, and if her influence was relatively lessened by the tendency of large to absorb small States, which had been one effect of the late disorder of Europe, she remained the France

Congress of
Vienna,
Sept., 1814,
March,
1815.

of Louis XVI. The Peace of Paris, as it was called, was followed by a Congress, to resettle the Continent, held at Vienna in the autumn of 1814; and at this great Council the Northern Powers exhibited an ambitious lust for dominion not unworthy of Napoleon himself. Russia threatened to swallow the whole of Poland; and Prussia, not contented with the enormous spoil she had acquired by taking part alternately with France and the allied Powers, aspired to annex a large part of Germany; and their

pretensions became so intolerable that a fresh general war seemed for a while imminent. Meantime
Louis XVIII., the new King of France, had endeavoured to consolidate his power ; but the difficulties in his way were, perhaps, invincible. The *Unpopu-*
larity of
Bourbon Monarchy was soon felt to represent na- *the Govern-*
ment of
tional disaster and disgrace ; if France had eagerly *Louis*
XVIII.
grasped at peace, she quickly learned to dislike
her position as a conquered Power not of the first
class, and she charged on her rulers the bitter consequences of humiliation, subjugation, and defeat.
The government of the King, too, made several
mistakes, and the associations which gathered round
it contributed to excite alarm and suspicion. The
old Imperial army was broken up, and deprived of
the far-famed Tricolour ; many of the new revolutionary interests were menaced, if not openly attacked ; invidious distinctions were drawn which
disturbed the civil equality won in 1789 ; and plans
were formed for changes which seemed to shake the
innumerable titles founded on the immense confiscations of bygone years. The general feeling of ill- *Conduct of*
the émi-
will was increased by the attitude and conduct of *grés.*
the surviving *émigrés* who had returned with Louis
from exile ; these representatives of a detested past,
who, it was bitterly said, ' could neither forget nor
learn,' talked loudly of restoring the feudal abuses,
and of taking their own in due time ; and, high
placed and caressed at court, they delighted to dis-

play towards the upstart *noblesse* of the 'Corsican monster,' as he was called, the refined insolence of an exclusive caste. The fine ladies of this worthy order of men were singularly skilful, as may be supposed, in this exhibition of the breeding of Versailles.

The general result of this state of things was that, within a few months after his elevation to the throne, France became hostile to her new monarch, and, filled with sullen jealousy and discontent, began to hope wistfully for some unknown change. The sentiment of irritation soon proved intense in the army still true to its mighty chief; and it was shared by the whole class of younger officers, though the ennobled marshals of the fallen Emperor felt or feigned respect for the restored dynasty. All this was not lost on the extraordinary man who, from his island speck in the Mediterranean, kept his eyes fixed on the state of Europe; and by degrees Na-

Napoleon leaves Elba. Feb. 26, May, 1815

poleon conceived the design of escaping from the kind of royal captivity in which he had been lately placed. His preparations were not long in being made, and on February 26, 1815, he set off on the most daring enterprise which even his sanguine mind had formed—that of recovering his lost Empire in the face, as it seemed, of all Europe against him. A few hundred men of the Imperial Guard, left about him incautiously by the Allies, accompanied the adventurer in a flotilla; and it is

but just to say that if his attempt was a breach of
faith as regards Europe, it was hardly so as regards
Louis XVIII., who had been intriguing against a still
feared rival. On the 1st of March the little expedition set foot on the shores of Provençe, not far
from the spot where years before the youthful Bonaparte had returned from Egypt; and the strange
apparition was soon welcomed with sentiments of
exultation and joy, for the neighbouring peasantry
had not forgotten how Marengo had freed them from
foreign invasion. In a few hours the exile was
threading his way through the defiles of Dauphiny,
issuing on his path proclamations appealing to
French patriotism; and his march before long began
to resemble the rapid spread of some mighty influence which, for the moment, nothing can resist.
Regiment after regiment, sent to check his progress,
threw down their arms at the well-known sight of
their loved and unforgotten commander; and in a
short time his insignificant band had gathered into
a considerable force, which multiplied at every
stage of his advance. He was at Grenoble on
March 9, and by the 10th had taken possession of
Lyons; and, as he moved onwards, hostile authority
seemed to disappear and perish before him. The
whole army was now in revolt; and Ney, one of his
most brilliant lieutenants, having been swept away
in the general torrent, the Bourbon cause soon became desperate, and Louis XVIII. fled across the

He lands in
France,
March 1,
1815.

His tri-
umphant
march to
Paris.

frontier. On March 20 the restored exile was once more in his place at the Tuileries; and, before a fortnight had passed, a faint show of royalist opposition had been quietly put down. Yet though in Napoleon's expressive phrase, 'his eagle had flown from steeple to steeple with the Tricolour to the towers of Notre Dame,' the Revolution which had reseated him on the throne was in the main the work of the army; and if France, fascinated, as it were, at the sight, seemed to welcome her returning master again, she rather rejoiced that the Bourbons were gone than believed or even hoped that the Empire could live.

Pacific overtures of Napoleon.

Napoleon, upon regaining the throne, assured the Great Powers of his desire for peace, and soon afterwards proceeded to offer a more liberal Constitution to France than she had possessed at any previous time, with a double Assembly, and guarantees for freedom. It is useless to enquire whether the Emperor was sincere; but it is not surprising that he was not believed, and he was quickly undeceived even if he imagined that he could play the part of a 'Napoleon of Peace.' At the intelli-

The Allied Powers declare war, March 25, 1815.

gence of his return from Elba, the discords of the Coalition ceased; and after proclaiming Napoleon an outlaw, the Great Powers set their armies in motion to crush the usurper and invade France again. Left thus to contend against Europe in arms, Napoleon tried to confront the approaching tempest; and, notwithstanding all that detractors have said, his efforts were great and worthy of him. He did not, indeed, appeal to the Nation, true to the last to his despotic instincts, or revive the memories of 1793, and France was still much too worn out to display the enthusiasm of that time ; but he effected all that ability could effect; and if he ultimately failed, it was because the nature of the present contest had nothing in common with that in which the Convention triumphed. One fortunate circumstance was in his favour ; many thousands of prisoners had returned home, and by making use of these old soldiers and turning to the best account the resources of France, he raised the French army from a state of impotence to a force of not less than 600,000 men, of whom 200,000 were ready for the field. Two strategic combinations were now before him : he might either await the attack of the Allies around Paris, which he had hastily fortified, or he might suddenly assume the offensive, and, falling upon one of their separate màsses, endeavour to divide and beat them in detail. Adhering to his usual system of war, he resolved to adopt the second plan ; and if possibly it was the less prudent, it was in some particulars extremely tempting. On the extreme end of the front of invasion on which the hosts of the Coalition would move, the two armies of Blücher and Wellington lay encamped in Belgium from the Scheldt to the Meuse ; and they were exposed to a fierce and sudden attack, as they were extended

Great efforts of Napoleon to restore the French army.

Campaign of 1815.

Two plans of operations open to Napoleon.

He resolves to attack Blücher and Wellington in Belgium.

Y

along the French frontier, and their supports were still on the Elbe and the Oder. It might be possible, thus, to assail and divide this detached wing of the hostile arrays, and to destroy successively its isolated parts ; and if a decisive victory were won, who could tell what the results would be ? And if the Emperor should be inferior in force, many a field of fame could attest that his genius had been able to turn the scales of fortune when placed in a position of this kind.

Concentration of the French army on the frontier.

In the second week of June the movement began on which the Emperor had staked his destiny. The French divisions, their movements concealed by false demonstrations with exquisite skill, drew together rapidly from Lille to Metz, while the Imperial Guard pressed forward from Paris, the Emperor's object being to combine his forces secretly and swoop on Belgium. Napoleon left the capital on June 12 ; and by the evening of the 14th his whole army, concentrated with extraordinary art, was collected on the edge of the French frontier, immediately around the banks of the Sambre. It numbered about 130,000 men; but though a sudden rising in La Vendée had deprived its chief of 20,000 more, and the united armies of Blücher and Wellington were fully 220,000 strong, Napoleon drew, from what he had already achieved, a hopeful augury of brilliant success. On the morning of the

It advances on June 15, 1815.

15th the march began, but though skilfully delayed by a Prussian detachment, the French columns ad-

vanced rapidly; and having passed the Sambre and
seized Charleroi, made straight for the centre of
the allied line, the great road from Namur to Brussels, which, as Napoleon calculated, was but weakly
defended. The French army, before night had
closed, lay between Gosselies, Frasne, and Fleurus;
and if it had not got quite so far as its leader had
wished, it was even now in a most formidable position, within easy reach of the advanced posts of its
foes, not as yet concentrated in adequate strength.
On the 16th the French advanced again; and
Blücher, who, with his wonted daring, was eager to
fight as soon as possible, offered battle to Napoleon
near Ligny, though his forces were not nearly collected, and Wellington had urged him not to run
the risk. The engagement was one of the fiercest
on record, each side contending with a national
hatred; but the skill of Napoleon at last triumphed;
and the Prussian army, pierced through the centre,
was driven with heavy loss from the field. Meanwhile Ney had attacked Wellington at Quatre Bras,
a few miles to the left; but though the British
chief could send no aid to Blücher, he held Ney in
check, and preserved the Prussians from an attack
on the flank designed by the Emperor, which would
have made Ligny a second Jena. An accident, however, alone prevented this consummation from being
otherwise attained. Ney had left a part of his forces
in his rear; and Napoleon having perceived from

Battle of Ligny, June 16, 1815.

Battle of Quatre Bras, June 16, 1815.

Y 2

Ligny that his lieutenant was making but little progress, he ordered this division to advance and accomplish the task of Ney, and complete the defeat of Blücher.   Ney, however, severely pressed by Wellington, called this detachment to him at the critical moment; and this misadventure probably had a decisive influence on the result of the campaign.

Result of the operations of June 16.

These operations had given the French a brilliant triumph over the Prussians, had brought them upon the allied centre, and had prevented Blücher and Wellington joining on what was their proper line of junction, the before-named road from Namur to Brussels. Still the Prussian army had not been routed as the Emperor had had good reason to hope ; and the allied chiefs might yet find the means of uniting by activity and zeal, an event which might lead to Napoleon's ruin. The Emperor, however, after Ligny, appears to have thought that, for some days at least, he had got rid of the defeated Prussians, and that he would have ample time to turn against Wellington ; and this conclusion would probably have been entirely correct in his earlier campaigns. Events, however, were soon to show what the energy of the allied chiefs and the passions which sustained the Prussians could effect. The Prussian army, though beaten at Ligny, had not

Blücher rallies the Prussians, and moves to join Wellington

been in the least cowed ; Blücher had rallied it with heroic vigour ; and he had soon concentrated his whole forces, and made them ready for a new effort, in positions only a few miles from Wellington.

The British commander prepared to approach his on a second line. colleague by a corresponding movement; and thus, though forced from the first line, the allied generals were not really divided, and were beginning to approach each other on a second. The French, meanwhile, had been allowed to halt, worn out by continued marches and fighting, nor had the movement of the retiring Prussians been watched and followed with sufficient care; and, accordingly, when about mid-day on the 17th, Napoleon broke up to assail Wellington, he had no conception that the Prussian army was not far off, and was drawing towards the British. He left Quatre Bras with about Movements of Napoleon and Wellington on June 17, 1815. 72,000 men, having detached Grouchy with 34,000 to 'observe the Prussians and complete their defeat;' but Wellington was already falling back; and by the evening he had taken a position beyond the little village of Waterloo, resolved to accept battle on a pledge from Blücher—who, at this time, had his whole army at Wavre, twelve miles away—that he would come up to assist the British. Meanwhile Miscalculations of Napoleon. Grouchy, who had completely lost sight of the Prussians, and even of the line of their march, and who, besides, like the Emperor, thought they could not yet venture to join Wellington, had advanced only a short way from Ligny; and, ignorant what dispositions to make, had halted in the neighbourhood of Gembloux, at a considerable distance in the rear of Napoleon, and separated from Blücher by no small interval.

Results of
the opera-
tions of
June 17.
By these arrangements it had been made all but
certain that the allied armies would unite at Waterloo in sufficient time to overpower the French; and
the chances were faint that Grouchy at Gembloux
would be able to arrest the march of Blücher. The
Emperor, however, either still convinced that the
Prussians were far away from the field, or that
Grouchy possessed the means to stop them, thought
only of bringing Wellington to bay; and as Wellington had only 69,000 men, composed in part of
second-rate troops, and was very inferior in horse
and guns, his adversary felt assured of victory.
Napoleon wished to attack at daybreak on the 18th;
but the night and morning had been dense with
rain, and he delayed the attack for several hours, in
order to allow the ground to harden, and to give his
manœuvres more effect—a sure proof that he had no
conception that Blücher was already gathering on
Great
battle of
Waterloo,
June 18,
1815.
his flank. The battle began by an assault on
Hougoumont, an advanced post on the British right;
but this was intended to be a feint; and it was succeeded by a tremendous onslaught on Wellington's
left and left centre, which met a brilliant and
decisive repulse. Meanwhile Napoleon had been
informed that about 30,000 men of Blücher's forces
had advanced from Wavre, and were close at
hand; and, accordingly, at about midday he sent
part of his reserve against this unexpected foe,
though he still hoped it was a stray column which
he would be able to hold in check. The plan of

Napoleon's battle was thus much disturbed; but he turned fiercely against the British centre; and, after a series of furious attacks, the French became masters of La Haye Sainte, a farm-house in front of Wellington's line. The violence of their efforts now became intense; the French cavalry streamed up the slopes of Mont St. Jean, and fell desperately on the British position; but nothing could break the infantry of the defence, which in solid squares ' seemed rooted to the earth;' and after a succession of fruitless charges the horsemen, who were unsupported by foot, were obliged, cruelly mutilated, to retreat. During all this time the Prussian detachment had being striking hardly at Napoleon's right; and this had given Wellington relief not sufficiently acknowledged by English writers; but about seven this attack seemed spent; and Napoleon seized the opportunity to make a last attempt against the British centre. The greater part of the Imperial Guard, the veterans of a hundred fields, marched resolutely to this fresh encounter; but Wellington had skilfully strengthened his line; and, after a short but terrible struggle, the Guard was repulsed and swayed slowly backward. It was now the turn of the British to advance; and just at this moment the remaining masses of Blücher appeared upon the field, and rending asunder the French right, converted defeat into a frightful rout. Except the Guard, which fought to the last, Napoleon's army became a mere chaos of despairing fugitives pursued

Defeat and rout of the French army.

by the Prussians; and only a fragment of the ruined host was ever seen in arms again. Grouchy, who had broken up from Gembloux late, and had refused, when urged, to approach Waterloo, only reached Wavre to find Blücher gone, and merely detained 15,000 Prussians from the scene where the Empire had succumbed.

Reflections on the campaign.   Volumes have been written on this memorable struggle, yet the general facts are sufficiently plain. The first operations of the French Emperor were a masterpiece of military skill; and the result was that, in spite of a very great preponderance of force, Blücher and Wellington were in peril on June 16, and probably, but for a mere accident, Ligny would have been an overwhelming defeat. The Emperor's movements after the 16th have been condemned by the worshippers of success; but all that can be fairly said is that he sanctioned certain errors of detail, for which a commander-in-chief can be scarcely blamed, and that he made a single false calculation, fatal in the event, but extremely natural. The delays of the French on the 17th should be ascribed to the fatigues of the troops; if the Prussians were not sufficiently watched, surely the fault lies mainly with the French staff; and as for the supposition that Blücher could not join Wellington for some days, Napoleon's views were warranted by his earlier campaigns, and had proved correct on similar occasions. It was in fact most unlikely that the defeated Prussians would be able to make a

critical march and fight at Waterloo on June 18;
and that such a movement became possible was
largely caused by a moral element—the passions that
stirred the army of Blücher.   Nor did Napoleon
neglect the Prussians; he detached Grouchy to hold
them in check; and the conduct of his lieutenant
was wretched, even if we may doubt that with
34,000 men he could have stopped Blücher with
90,000.   Napoleon was not a ' mere shadow of his
former self' in 1815; and if he met ruin on the
field of Waterloo, it was not because his powers had
declined, but that—apart from the over-confidence
which we see in this as in other campaigns—his
antagonists supported each other better than any
allied chiefs had ever done before, and especially
that the Prussian army, sustained by a principle he
undervalued, baffled reasoning, founded on experience, indeed, but fatally untrue in the actual
contest.   If this view be right, the defeat of Napoleon was largely due to his characteristic contempt of some of the strongest feelings that animate
man; and the frequent errors of the politician con-
founded the schemes of the military chief.   As for
the conduct of the allied commanders, it exposed
them to danger at the outset; and as Blücher ought
not to have fought at Ligny, it revealed at first the
divided councils so often disastrous to allies.   But
all this was nobly repaired; and the constancy of
Wellington on the field of Waterloo, and the heroism

of Blücher in overcoming defeat, are fine specimens of great qualities. Yet though Waterloo was a splendid triumph, the fame of Wellington does not rest on the campaign of 1815 as a whole; his real title to renown depends on the admirable sagacity with which he perceived the weak point in Napoleon's strategy, and illustrated a discovery, big with great results, by his memorable defence of Torres Vedras.

Conclusion. Napoleon abdicated after the rout of Waterloo, the French Chambers, already hostile, rising against him in the hour of disaster; and before long he was on his way to the last scene of his eventful history, the solitary island of St. Helena. France, trodden under foot by the allied hosts, accepted the Bourbons in 1815, as she had accepted them the year before; but though Louis XVIII. was a sagacious ruler, such a dynasty could not become permanent. A sudden heave of the revolutionary forces which, though long quiescent, retained life, deprived Charles X. of his crown; and a Constitutional Monarchy was set up in his stead, in favour of the son of the Duke of Orleans, the Royal Jacobin of 1793. This government, of which the chief feature was a corrupt and weak parliamentary system, met the fate of its immediate forerunner, and it was followed by a short-lived Republic, which, after agitating Europe in 1848, perished unlamented in 1851. Long before this time the great name of Napoleon

had regained its magical power in France, and the
nephew of the departed conqueror, a grandson of the
divorced Josephine, was raised to the throne as
Emperor of the French, assuming the title of Napoleon III. The Second Empire was a feeble image
of the first, without the military genius of its chief;
and it disappeared in the great war of 1870, in
which Prussia, heading a united Germany, more
than avenged the disaster of Jena, and has torn
from France Alsace and Lorraine, spared in 1814
and 1815. A provisional Republic has been since in
power, its history marked by a national defence as
gallant as that of 1793, but less noticed because a
failure, and by a terrific outbreak of Jacobin frenzy
which awed Europe in 1871; but this settlement is
felt to be only for a time; and France remains torn
by revolutionary troubles kept under only by the
power of the sword in the hands of a soldier brave
indeed, but not a chief of the first order. The
general results of these events, which all run up to
1789–1815, are that Government in France is never
secure, and that the nation appears to have lost
some essential elements of general welfare; and
though the great convulsion of the last century is
not the only, it certainly is a principal cause of this
evil disorder. If the material progress of France,
too, since the fall of Feudalism has been immense,
there has been no corresponding moral improvement;
and if, within the memory of living man, she swayed

Europe from the Tagus to the Baltic, her military reverses have since that time been awful, and the Tricolour has been plucked down from Metz and Strasburg, which once floated on Madrid and Moscow. The consequences of the Revolution outside France have been, on the whole, more fruitful of good; they have tended to civilisation and national progress, but they have been accompanied all over Europe by frightful wars and general disturbance; and we see the evils in the prodigious armaments and fierce animosities of the Continent, and in the disregard of the rights of the weak, and the ignoble flattery of force and success, too characteristic of modern politics. We end as we began; it is at least doubtful whether the mischief done by the French Revolution does not preponderate over its benefits. The greatest of English historians remarked, a few years before 1789, that the era of wars seemed about to close, and that Europe would be for all time secure from the barbarism of the savage hordes which had overturned Imperial Rome. What would Gibbon have said had he lived to witness Borodino, Leipsic, Waterloo, Sedan, and the atrocities of the Reign of Terror, and of the Commune of Paris in 1871!

LONDON: PRINTED BY
SPOTTISWOODE AND CO., NEW-STREET SQUARE
AND PARLIAMENT STREET

GENERAL LIST OF WORKS

PUBLISHED BY

Messrs. LONGMANS, GREEN, AND CO.

PATERNOSTER ROW, LONDON.

————∘∘⊱✠⊰∘∘————

History, Politics, Historical Memoirs, &c.

The **HISTORY of ENGLAND** from the Fall of Wolsey to the Defeat of the Spanish Armada. By JAMES ANTHONY FROUDE, M.A. late Fellow of Exeter College, Oxford.

> LIBRARY EDITION, Twelve Volumes, 8vo. price £8. 18s.
> CABINET EDITION, Twelve Volumes, crown 8vo. price 72s.

The **ENGLISH in IRELAND** in the **EIGHTEENTH CENTURY.** By JAMES ANTHONY FROUDE, M.A. late Fellow of Exeter College, Oxford. 3 vols. 8vo. price 48s.

ESTIMATES of the **ENGLISH KINGS** from **WILLIAM** the **CON-QUEROR** to GEORGE III. By J. LANGTON SANFORD. Crown 8vo. 12s. 6d.

The **HISTORY of ENGLAND** from the Accession of James II. By Lord MACAULAY.

> STUDENT'S EDITION, 2 vols. crown 8vo. 12s.
> PEOPLE'S EDITION, 4 vols. crown 8vo. 16s.
> CABINET EDITION, 8 vols. post 8vo. 48s.
> LIBRARY EDITION, 5 vols. 8vo. £4.

LORD MACAULAY'S WORKS. Complete and Uniform Library Edition. Edited by his Sister, Lady TREVELYAN. 8 vols. 8vo. with Portrait, price £5. 5s. cloth, or £8. 8s. bound in tree-calf by Rivière.

On **PARLIAMENTARY GOVERNMENT** in **ENGLAND**; its Origin, Development, and Practical Operation. By ALPHEUS TODD, Librarian of the Legislative Assembly of Canada. 2 vols. 8vo. price £1. 17s.

The **CONSTITUTIONAL HISTORY** of **ENGLAND,** since the Accession of George III. 1760—1860. By Sir THOMAS MAY, C.B. The Fourth Edition, thoroughly revised. 3 vols. crown 8vo. price 18s.

DEMOCRACY in **EUROPE**; a History. By Sir THOMAS ERSKINE MAY, K.C.B. 2 vols. 8vo. *[In the press.*

The **HISTORY of ENGLAND,** from the Earliest Times to the Year 1865. By C. D. YONGE, B.A. Second Edition. Crown 8vo. 7s. 6d.

The **ENGLISH GOVERNMENT** and **CONSTITUTION** from Henry VII. to the Present Time. By JOHN Earl RUSSELL, K.G. Fcp. 8vo. 3s. 6d.

A

The **OXFORD REFORMERS** — John Colet, Erasmus, and Thomas More ; being a History of their Fellow-work. By FREDERIC SEEBOHM. Second Edition, enlarged. 8vo. 14s.

LECTURES on the **HISTORY** of **ENGLAND**, from the Earliest Times to the Death of King Edward II. By WILLIAM LONGMAN. With Maps and Illustrations. 8vo. 15s.

The **HISTORY** of the **LIFE** and **TIMES** of **EDWARD** the **THIRD.** By WILLIAM LONGMAN. With 9 Maps, 8 Plates, and 16 Woodcuts. 2 vols. 8vo. 28s.

HISTORY of **MARY STUART QUEEN** of **SCOTS.** Translated from the Original MS. of Professor PETIT. By C. DE FLANDRE, F.S.A. Scot. Professor of the French Language and Literature in Edinburgh. With two Portraits. 2 vols. 4to. 63s.

WATERLOO LECTURES ; a Study of the Campaign of 1815. By Colonel CHARLES C. CHESNEY, R.E. New Edition. 8vo. with Map, 10s. 6d.

The **LIFE and TIMES** of **SIXTUS** the **FIFTH.** By Baron HÜBNER. Translated with the Author's sanction, by H. E. H. JERNINGHAM. 2 vols. 8vo. 24s.

The **SIXTH ORIENTAL MONARCHY**; or, the Geography, History, and Antiquities of Parthia. By GEORGE RAWLINSON, M.A. Professor of Ancient History in the University of Oxford. Maps and Illustrations. 8vo. 16s.

The **SEVENTH GREAT ORIENTAL MONARCHY**; or, a History of the Sassanians : with Notices, Geographical and Antiquarian. By G. RAWLINSON, M.A. Professor of Ancient History in the University of Oxford. 8vo. with Maps and Illustrations. *[In the press.*

A HISTORY of **GREECE.** By the Rev. GEORGE W. COX, M.A. late Scholar of Trinity College, Oxford. VOLS. I. & II. (to the Close of the Peloponnesian War) 8vo. with Maps and Plans, 36s.

The **HISTORY OF GREECE.** By C. THIRLWALL, D.D. Lord Bishop of St. David's. 8 vols. fcp. 8vo. 28s.

GREEK HISTORY from Themistocles to Alexander, in a Series of Lives from Plutarch. Revised and arranged by A. H. CLOUGH. New Edition. Fcp. with 44 Woodcuts, 6s.

The **TALE** of the **GREAT PERSIAN WAR**, from the Histories of Herodotus. By GEORGE W. COX, M.A. New Edition. Fcp. 3s. 6d.

The **HISTORY** of **ROME.** By WILLIAM IHNE. English Edition, translated and revised by the Author. VOLS. I. and II. 8vo. price 30s.

HISTORY of the **ROMANS** under the **EMPIRE.** By the Very Rev. C. MERIVALE, D.C.L. Dean of Ely. 8 vols. post 8vo. 48s.

The **FALL** of the **ROMAN REPUBLIC**; a Short History of the Last Century of the Commonwealth. By the same Author. 12mo. 7s. 6d.

THREE CENTURIES of **MODERN HISTORY.** By CHARLES DUKE YONGE, B.A. Crown 8vo. 7s. 6d.

The **STUDENT'S MANUAL** of the **HISTORY** of **INDIA**, from the Earliest Period to the Present. By Colonel MEADOWS TAYLOR, M.R.A.S. M.R.I.A. Second Thousand. Crown 8vo. with Maps, 7s. 6d.

The **HISTORY** of **INDIA**, from the Earliest Period to the close of Lord Dalhousie's Administration. By J. C. MARSHMAN. 3 vols. crown 8vo. 22s. 6d.

INDIAN POLITY: a View of the System of Administration in India. By Lieutenant-Colonel GEORGE CHESNEY, Fellow of the University of Calcutta. New Edition, revised; with Map. 8vo. price 21s.

The **IMPERIAL** and **COLONIAL CONSTITUTIONS** of the **BRITANNIC EMPIRE**, including INDIAN INSTITUTIONS. By Sir EDWARD CREASY, M.A. With 6 Maps. 8vo. price 15s.

The **HISTORY** of **PERSIA** and its **PRESENT POLITICAL SITUATION**; with Abstracts of all Treaties and Conventions between Persia and England, and of the Convention with Baron Reuter. By CLEMENTS R. MARKHAM, C.B. F.R.S. 8vo. with Map, 21s.

REALITIES of **IRISH LIFE**. By W. STEUART TRENCH, late Land Agent in Ireland to the Marquess of Lansdowne, the Marquess of Bath, and Lord Digby. Cheaper Edition. Crown 8vo. price 2s. 6d.

The **STUDENT'S MANUAL** of the **HISTORY** of **IRELAND**. By MARY F. CUSACK. Crown 8vo. price 6s.

CRITICAL and **HISTORICAL ESSAYS** contributed to the *Edinburgh Review*. By the Right Hon. LORD MACAULAY.

CABINET EDITION, 4 vols. post 8vo. 24s. | LIBRARY EDITION, 3 vols. 8vo. 36s.
PEOPLE'S EDITION, 2 vols. crown 8vo. 8s. | STUDENT'S EDITION, 1 vol. cr. 8vo. 6s.

HISTORY of **EUROPEAN MORALS**, from Augustus to Charlemagne By W. E. H. LECKY, M.A. Second Edition. 2 vols. 8vo. price 28s.

HISTORY of the **RISE** and **INFLUENCE** of the **SPIRIT** of RATIONALISM in EUROPE. By W. E. H. LECKY, M.A. Cabinet Edition, being the Fourth. 2 vols. crown 8vo. price 16s.

The **HISTORY** of **PHILOSOPHY**, from Thales to Comte. By GEORGE HENRY LEWES. Fourth Edition. 2 vols. 8vo. 32s.

The **HISTORY** of the **PELOPONNESIAN WAR**. By THUCYDIDES. Translated by R. CRAWLEY, Fellow of Worcester College, and formerly Scholar of University College, Oxford. 8vo. *[In the press.*

The **MYTHOLOGY** of the **ARYAN NATIONS**. By GEORGE W. COX, M.A. late Scholar of Trinity College, Oxford, 2 vols. 8vo. 28s.

HISTORY of **CIVILISATION** in England and France, Spain and Scotland. By HENRY THOMAS BUCKLE. New Edition of the entire Work, with a complete INDEX. 3 vols. crown 8vo. 24s.

HISTORY of the **CATHOLIC CHURCH** of **JESUS CHRIST** from the Death of St. John to the Middle of the Second Century. By the Rev. T. W. MOSSMAN, B.A. 8vo. price 16s.

HISTORY of the **CHRISTIAN CHURCH**, from the Ascension of Christ to the Conversion of Constantine. By E. BURTON, D.D. late Prof. of Divinity in the Univ. of Oxford. New Edition. Fcp. 3s. 6d.

SKETCH of the **HISTORY** of the **CHURCH** of **ENGLAND** to the Revolution of 1688. By the Right Rev. T. V. SHORT, D.D. Lord Bishop of St. Asaph. Eighth Edition. Crown 8vo. 7s. 6d.

HISTORY of the **EARLY CHURCH**, from the First Preaching of the Gospel to the Council of Nicæa, A.D. 325. By Miss SEWELL. Fcp. 8vo. 4s. 6d.

MAUNDER'S HISTORICAL TREASURY; comprising a General Introductory Outline of Universal History, and a series of Separate Histories. Latest Edition, revised and brought down to the Present Time by the Rev. GEORGE WILLIAM COX, M.A. Fcp. 8vo. 6s. cloth, or 10s. calf.

A 2

CATES' and WOODWARD'S ENCYCLOPÆDIA of CHRONOLOGY,
HISTORICAL and BIOGRAPHICAL; comprising the Dates of all the Great
Events of History, including Treaties, Alliances, Wars, Battles, &c.; Incidents
in the Lives of Eminent Men and their Works, Scientific and Geographical Discoveries, Mechanical Inventions, and Social Improvements. 8vo. price 42s.

The FRENCH REVOLUTION and FIRST EMPIRE; an Historical
Sketch. By WILLIAM O'CONNOR MORRIS, sometime Scholar of Oriel College,
Oxford. Post 8vo.   [Nearly ready.

The HISTORICAL GEOGRAPHY of EUROPE. By E. A. FREEMAN,
D.C.L. late Fellow of Trinity College, Oxford. 8vo. Maps.   [In the press.

EPOCHS of HISTORY: a Series of Books treating of the History of
England and Europe at successive Epochs subsequent to the Christian Era.
Edited by EDWARD F. MORRIS, M.A. of Lincoln College, Oxford. In fcp. 8vo.
volumes of about 230 pages each. The three following are advancing at
press:—

> **The Crusades.** By the Rev. G. W. Cox, M.A. late Scholar of Trinity
> College, Oxford.
>
> **The Era of the Protestant Revolution.** By F. SEEBOHM.
>
> **The Thirty Years' War, 1618-1648.** By SAMUEL RAWSON GARDINER,
> late Student of Christ Church.

Biographical Works.

AUTOBIOGRAPHY. By JOHN STUART MILL. 8vo. price 7s. 6d.

The LIFE of NAPOLEON III. derived from State Records, Unpublished
Family Correspondence, and Personal Testimony. By BLANCHARD JERROLD.
4 vols. 8vo. with Portraits from the Originals in possession of the Imperial
Family, and Facsimiles of Letters of Napoleon I. Napoleon III. Queen
Hortense, &c.   [VOL. I. nearly ready.

LIFE and LETTERS of Sir GILBERT ELLIOT, First EARL of
MINTO, from 1751 to 1806, when his Public Life in Europe was closed by his
Appointment as the Vice-Royalty of India. Edited by his Grand-Niece, the
COUNTESS of MINTO. 3 vols. 8vo. 31s. 6d.

MEMOIR of THOMAS FIRST LORD DENMAN, formerly Lord Chief
Justice of England. By Sir JOSEPH ARNOULD, B.A. K.B. late Judge of the High
Court of Bombay. With 2 Portraits. 2 vols. 8vo. 32s.

ESSAYS in MODERN MILITARY BIOGRAPHY. By CHARLES
CORNWALLIS CHESNEY, Lieutenant-Colonel in the Royal Engineers. 8vo. 12s. 6d.

BIBLIOTHECA CORNUBIENSIS; a Catalogue of the Writings, both
MS. and printed, of Cornishmen from the Earliest Times, and of Works relating
to the County of Cornwall. With Biographical Memoranda and copious Literary
References. By G. C. BOASE and W. P. COURTNEY. In Two Volumes. VOL. I.
A.—O. Imperial 8vo. 21s.

SHAKESPEARE'S HOME and RURAL LIFE. By JAMES WALTER,
Major 4th Lancashire Artillery Volunteers. Comprising a Biographical Narrative,
illustrated by about 100 Landscapes and Views produced by the Heliotype process from Photographs taken in the localities. Imperial 4to. 52s. 6d.

BIOGRAPHICAL and CRITICAL ESSAYS, reprinted from Reviews,
with Additions and Corrections. Second Edition of the Second Series. By A.
HAYWARD, Q.C. 2 vols. 8vo. price 28s. THIRD SERIES, in 1 vol. 8vo. price 14s.

The **LIFE of LLOYD, FIRST LORD KENYON, LORD CHIEF JUSTICE of ENGLAND.** By the Hon. GEORGE T. KENYON, M.A. of Ch. Ch. Oxford. With Portraits. 8vo. price 14s.

MEMOIR of GEORGE EDWARD LYNCH COTTON, D.D. Bishop of Calcutta and Metropolitan. With Selections from his Journals and Correspondence. Edited by Mrs. COTTON. Crown 8vo. 7s. 6d.

MEMOIR of the LIFE of Admiral Sir EDWARD CODRINGTON. With Selections from his Public and Private Correspondence. Edited by his Daughter, Lady BOURCHIER. Portraits, Maps, and Plans. 2 vols. 8vo. 36s.

LIFE of ALEXANDER VON HUMBOLDT. Compiled in Commemoration of the Centenary of his Birth, and edited by Professor KARL BRUHNS; translated by JANE and CAROLINE LASSELL, with 3 Portraits. 2 vols. 8vo. 36s.

MEMOIRS of BARON STOCKMAR. By his SON, Baron E. VON STOCKMAR. Translated from the German by G. A. M. Edited by F. MAX MÜLLER, M.A. 2 vols. crown 8vo. price 21s.

LORD GEORGE BENTINCK; A Political Biography. By the Right Hon. BENJAMIN DISRAELI, M.P. Crown 8vo. price 6s.

The **LIFE OF ISAMBARD KINGDOM BRUNEL, Civil Engineer.** By ISAMBARD BRUNEL, B.C.L. With Portrait, Plates, and Woodcuts. 8vo. 21s.

RECOLLECTIONS of PAST LIFE. By Sir HENRY HOLLAND, Bart. M.D. F.R.S. late Physician-in-Ordinary to the Queen. Third Edition. Post 8vo. price 10s. 6d.

The **LIFE and LETTERS of the Rev. SYDNEY SMITH.** Edited by his Daughter, Lady HOLLAND, and Mrs. AUSTIN. Crown 8vo. price 6s.

LEADERS of PUBLIC OPINION in IRELAND; Swift, Flood, Grattan, and O'Connell. By W. E. H. LECKY, M.A. New Edition, revised and enlarged. Crown 8vo. price 7s. 6d.

DICTIONARY of GENERAL BIOGRAPHY; containing Concise Memoirs and Notices of the most Eminent Persons of all Countries, from the Earliest Ages to the Present Time. Edited by W. L. R. CATES. 8vo. 21s.

LIVES of the QUEENS of ENGLAND. By AGNES STRICKLAND. Library Edition, newly revised; with Portraits of every Queen, Autographs and Vignettes. 8 vols. post 8vo. 7s. 6d. each.

LIFE of the DUKE of WELLINGTON. By the Rev. G. R. GLEIG, M.A. Popular Edition, carefully revised; with copious Additions. Crown 8vo. with Portrait, 5s.

FELIX MENDELSSOHN'S LETTERS from *Italy and Switzerland*, and *Letters from 1833 to 1847*, translated by Lady WALLACE. New Edition, with Portrait. 2 vols. crown 8vo. 5s. each.

MEMOIRS of SIR HENRY HAVELOCK, K.C.B. By JOHN CLARK MARSHMAN. Cabinet Edition, with Portrait. Crown 8vo. price 3s. 6d.

VICISSITUDES of FAMILIES. By Sir J. BERNARD BURKE, C.B. Ulster King of Arms. New Edition, remodelled and enlarged. 2 vols. crown 8vo. 21s.

The **RISE of GREAT FAMILIES,** other Essays and Stories. By Sir J. BERNARD BURKE, C.B. Ulster King of Arms. Crown 8vo. price 12s. 6d.

ESSAYS in ECCLESIASTICAL BIOGRAPHY. By the Right Hon.
Sir J. STEPHEN, LL.D. Cabinet Edition. Crown 8vo. 7s. 6d.

MAUNDER'S BIOGRAPHICAL TREASURY. Latest Edition, re-
constructed, thoroughly revised, and in great part rewritten; with 1,000 additional Memoirs and Notices, by W. L. R. CATES. Fcp. 8vo. 6s. cloth; 10s. calf.

LETTERS and LIFE of FRANCIS BACON, including all his Occa-
sional Works. Collected and edited, with a Commentary, by J. SPEDDING,
Trin. Coll. Cantab. 6 vols. 8vo. £3. 12s. Vol. VII. completion, nearly ready.

Criticism, Philosophy, Polity, &c.

A SYSTEMATIC VIEW of the SCIENCE of JURISPRUDENCE.
By SHELDON AMOS, M.A. Professor of Jurisprudence to the Inns of Court,
London. 8vo. price 18s.

A PRIMER of the ENGLISH CONSTITUTION and GOVERNMENT.
By SHELDON AMOS, M.A. Professor of Jurisprudence to the Inns of Court. New
Edition, revised. Post 8vo. [In the press.

The INSTITUTES of JUSTINIAN; with English Introduction, Translation and Notes. By T. C. SANDARS, M.A. New Edition. 8vo. 15s.

SOCRATES and the SOCRATIC SCHOOLS. Translated from the
German of Dr. E. ZELLER, with the Author's approval, by the Rev. OSWALD J.
REICHEL, M.A. Crown 8vo. 8s. 6d.

The STOICS, EPICUREANS, and SCEPTICS. Translated from the
German of Dr. E. ZELLER, with the Author's approval, by OSWALD J. REICHEL,
M.A. Crown 8vo. price 14s.

The ETHICS of ARISTOTLE, illustrated with Essays and Notes.
By Sir A. GRANT, Bart. M.A. LL.D. Third Edition, revised and partly
rewritten. [In the press.

The POLITICS of ARISTOTLE; Greek Text, with English Notes. By
RICHARD CONGREVE, M.A. New Edition, revised. 8vo. [Nearly ready.

The NICOMACHEAN ETHICS of ARISTOTLE newly translated into
English. By R. WILLIAMS, B.A. Fellow and late Lecturer of Merton College,
and sometime Student of Christ Church, Oxford. 8vo. 12s.

ELEMENTS of LOGIC. By R. WHATELY, D.D. late Archbishop of
Dublin. New Edition. 8vo. 10s. 6d. crown 8vo. 4s. 6d.

Elements of Rhetoric. By the same Author. New Edition. 8vo.
10s. 6d. crown 8vo. 4s. 6d.

English Synonymes. By E. JANE WHATELY. Edited by Archbishop
WHATELY. Fifth Edition. Fcp. 8vo. price 3s.

BACON'S ESSAYS with ANNOTATIONS. By R. WHATELY, D.D.
late Archbishop of Dublin. New Edition, 8vo. price 10s. 6d.

LORD BACON'S WORKS, collected and edited by J. SPEDDING, M.A.
R. L. ELLIS, M.A. and D. D. HEATH. 7 vols. 8vo. price £3. 13s. 6d.

ESSAYS CRITICAL and NARRATIVE, partly original and partly
reprinted from the Edinburgh, Quarterly, and other Reviews. By WILLIAM
FORSYTH, Q.C. M.P. for Marylebone. 8vo. [Now ready.

The **SUBJECTION** of **WOMEN.** By JOHN STUART MILL. New
Edition. Post 8vo. 5s.

On **REPRESENTATIVE GOVERNMENT.** By JOHN STUART MILL.
Crown 8vo. price 2s.

On **LIBERTY.** By JOHN STUART MILL. New Edition. Post
8vo. 7s. 6d. Crown 8vo. price 1s. 4d.

PRINCIPLES of **POLITICAL ECONOMY.** By the same Author.
Seventh Edition. 2 vols. 8vo. 30s. Or in 1 vol. crown 8vo. price 5s.

ESSAYS on **SOME UNSETTLED QUESTIONS** of **POLITICAL**
ECONOMY. By JOHN STUART MILL. Second Edition. 8vo. 6s. 6d.

UTILITARIANISM. By JOHN STUART MILL. New Edition. 8vo. 5s.

DISSERTATIONS and **DISCUSSIONS, POLITICAL, PHILOSOPHI-**
CAL, and HISTORICAL. By JOHN STUART MILL. 3 vols. 8vo. 36s.

EXAMINATION of Sir. W. **HAMILTON'S PHILOSOPHY,** and of the
Principal Philosophical Questions discussed in his Writings. By JOHN STUART
MILL. Fourth Edition. 8vo. 16s.

An **OUTLINE** of the **NECESSARY LAWS** of **THOUGHT;** a Treatise
on Pure and Applied Logic. By the Most Rev. W. THOMSON, Lord Archbishop
of York, D.D. F.R.S. Ninth Thousand. Crown 8vo. price 5s. 6d.

PRINCIPLES of **ECONOMICAL PHILOSOPHY.** By HENRY DUNNING
MACLEOD, M.A. Barrister-at-Law. Second Edition. In Two Volumes. VOL. I.
8vo. price 15s.

A **SYSTEM** of **LOGIC, RATIOCINATIVE** and **INDUCTIVE.** By JOHN
STUART MILL. Eighth Edition. Two vols. 8vo. 25s.

The **ELECTION** of **REPRESENTATIVES,** Parliamentary and Municipal; a Treatise. By THOMAS HARE, Barrister-at-Law. Crown 8vo. 7s.

SPEECHES of the **RIGHT HON. LORD MACAULAY,** corrected by
Himself. People's Edition, crown 8vo. 3s. 6d.

Lord Macaulay's Speeches on Parliamentary Reform in 1831 and
1832. 16mo. 1s.

FAMILIES of **SPEECH**: Four Lectures delivered before the Royal
Institution of Great Britain. By the Rev. F. W. FARRAR, D.D. F.R.S. New
Edition. Crown 8vo. 3s. 6d.

CHAPTERS on **LANGUAGE.** By the Rev. F. W. FARRAR, D.D. F.R.S.
New Edition. Crown 8vo, 5s.

A **DICTIONARY** of the **ENGLISH LANGUAGE.** By R. G. LATHAM,
M.A. M.D. F.R.S. Founded on the Dictionary of Dr. SAMUEL JOHNSON, as
edited by the Rev. H. J. TODD, with numerous Emendations and Additions.
In Four Volumes, 4to. price £7.

A **PRACTICAL ENGLISH DICTIONARY,** on the Plan of White's
English-Latin and Latin-English Dictionaries. By JOHN T. WHITE, D.D. Oxon.
and T. C. DONKIN, M.A. Assistant-Master, King Edward's Grammar School,
Birmingham. Post 8vo. 　　　　　　　　　　　　　　　　[In the press.

THESAURUS of **ENGLISH WORDS** and **PHRASES,** classified and
arranged so as to facilitate the Expression of Ideas, and assist in Literary
Composition. By P. M. ROGET, M.D. New Edition. Crown 8vo. 10s. 6d.

LECTURES on the SCIENCE of LANGUAGE. By F. MAX MÜLLER, M.A. &c. Seventh Edition. 2 vols. crown 8vo. 16s.

MANUAL of ENGLISH LITERATURE, Historical and Critical. By THOMAS ARNOLD, M.A. New Edition. Crown 8vo. 7s. 6d.

THREE CENTURIES of ENGLISH LITERATURE. By CHARLES DUKE YONGE. Crown 8vo. price 7s. 6d.

SOUTHEY'S DOCTOR, complete in One Volume. Edited by the Rev. J. W. WARTER, B.D. Square crown 8vo. 12s. 6d.

HISTORICAL and CRITICAL COMMENTARY on the OLD TESTA-MENT; with a New Translation. By M. M. KALISCH, Ph.D. VOL. I. *Genesis,* 8vo. 18s. or adapted for the General Reader, 12s. VOL. II. *Exodus,* 15s. or adapted for the General Reader, 12s. VOL. III. *Leviticus,* PART I. 15s. or adapted for the General Reader, 8s. VOL. IV. *Leviticus,* PART II. 15s. or adapted for the General Reader, 8s.

A DICTIONARY of ROMAN and GREEK ANTIQUITIES, with about Two Thousand Engravings on Wood from Ancient Originals, illustrative of the Industrial Arts and Social Life of the Greeks and Romans. By A. RICH, B.A. Third Edition, revised and improved. Crown 8vo. price 7s. 6d.

A LATIN-ENGLISH DICTIONARY. By JOHN T. WHITE, D.D. Oxon. and J. E. RIDDLE, M.A. Oxon. Revised Edition. 2 vols. 4to. 42s.

WHITE'S COLLEGE LATIN-ENGLISH DICTIONARY (Intermediate Size), abridged for the use of University Students from the Parent Work (as above). Medium 8vo. 18s.

WHITE'S JUNIOR STUDENT'S COMPLETE LATIN-ENGLISH and ENGLISH-LATIN DICTIONARY. New Edition. Square 12mo. price 12s.

Separately { The ENGLISH-LATIN DICTIONARY, price 5s. 6d.
The LATIN-ENGLISH DICTIONARY, price 7s. 6d.

A LATIN-ENGLISH DICTIONARY, for Middle-Class Schools, abridged from the Junior Student's Latin-English Dictionary. By JOHN T. WHITE, D.D. Oxon. 18mo. [*In the press.*

An ENGLISH-GREEK LEXICON, containing all the Greek Words used by Writers of good authority. By C. D. YONGE, B.A. New Edition. 4to. price 21s.

Mr. YONGE'S NEW LEXICON, English and Greek, abridged from his larger work (as above). Revised Edition. Square 12mo. price 8s. 6d.

A GREEK-ENGLISH LEXICON. Compiled by H. G. LIDDELL, D.D. Dean of Christ Church, and R. SCOTT, D.D. Dean of Rochester. Sixth Edition. Crown 4to. price 36s.

A Lexicon, Greek and English, abridged from LIDDELL and SCOTT's *Greek-English Lexicon.* Fourteenth Edition. Square 12mo. 7s. 6d.

A SANSKRIT-ENGLISH DICTIONARY, the Sanskrit words printed both in the original Devanagari and in Roman Letters. Compiled by T. BENFEY, Prof. in the Univ. of Göttingen. 8vo. 52s. 6d.

A PRACTICAL DICTIONARY of the FRENCH and ENGLISH LAN-GUAGES. By L. CONTANSEAU. Revised Edition. Post 8vo. 10s. 6d.

Contanseau's Pocket Dictionary, French and English, abridged from the above by the Author. New Edition, revised. Square 18mo. 3s. 6d.

NEW PRACTICAL DICTIONARY of the **GERMAN LANGUAGE;** German-English and English-German. By the Rev. W. L. BLACKLEY, M.A and Dr. CARL MARTIN FRIEDLÄNDER. Post 8vo. 7s. 6d.

The **MASTERY** of **LANGUAGES**; or, the Art of Speaking Foreign Tongues Idiomatically. By THOMAS PRENDERGAST. 8vo. 6s.

Miscellaneous Works and Popular Metaphysics.

ESSAYS on **FREETHINKING** and **PLAIN-SPEAKING.** By LESLIE STEPHEN. Crown 8vo. 10s. 6d.

MISCELLANEOUS and **POSTHUMOUS WORKS** of the Late **HENRY THOMAS BUCKLE.** Edited, with a Biographical Notice, by HELEN TAYLOR. 3 vols. 8vo. price 52s. 6d.

MISCELLANEOUS WRITINGS of **JOHN CONINGTON, M.A.** late Corpus Professor of Latin in the University of Oxford. Edited by J. A. SYMONDS, M.A. With a Memoir by H. J. S. SMITH, M.A. 2 vols. 8vo. 28s.

SEASIDE MUSINGS ON SUNDAYS AND WEEK-DAYS. By A. K. H. B. Crown 8vo. price 3s. 6d.

Recreations of a Country Parson. By A. K. H. B. FIRST and SECOND SERIES, crown 8vo. 3s. 6d. each.

The Common-place Philosopher in Town and Country. By A. K. H. B. Crown 8vo. price 3s. 6d.

Leisure Hours in Town; Essays Consolatory, Æsthetical, Moral, Social, and Domestic. By A. K. H. B. Crown 8vo. 3s. 6d.

The Autumn Holidays of a Country Parson; Essays contributed to *Fraser's Magazine*, &c. By A. K. H. B. Crown 8vo. 3s. 6d.

The Graver Thoughts of a Country Parson. By A. K. H. B. FIRST and SECOND SERIES, crown 8vo. 3s. 6d. each.

Critical Essays of a Country Parson, selected from Essays contributed to *Fraser's Magazine.* By A. K. H. B. Crown 8vo. 3s. 6d.

Sunday Afternoons at the Parish Church of a Scottish University City. By A. K. H. B. Crown 8vo. 3s. 6d.

Lessons of Middle Age; with some Account of various Cities and Men. By A. K. H. B. Crown 8vo. 3s. 6d.

Counsel and Comfort spoken from a City Pulpit. By A. K. H. B. Crown 8vo. price 3s. 6d.

Changed Aspects of Unchanged Truths; Memorials of St. Andrews Sundays. By A. K. H. B. Crown 8vo. 3s. 6d.

Present-day Thoughts; Memorials of St. Andrews Sundays. By A. K. H. B. Crown 8vo. 3s. 6d.

SHORT STUDIES on **GREAT SUBJECTS.** By JAMES ANTHONY FROUDE, M.A. late Fellow of Exeter Coll. Oxford. 2 vols. crown 8vo. price 12s.

LORD MACAULAY'S MISCELLANEOUS WRITINGS :—

LIBRARY EDITION. 2 vols. 8vo. Portrait, 21s.
PEOPLE'S EDITION. 1 vol. crown 8vo. 4s. 6d.

LORD MACAULAY'S MISCELLANEOUS WRITINGS and SPEECHES.
STUDENT'S EDITION, in crown 8vo. price 6s.

The Rev. SYDNEY SMITH'S ESSAYS contributed to the Edinburgh
Review. Authorised Edition, complete in 1 vol. Crown 8vo. price 2s. 6d. sewed
or 3s. 6d. cloth.

The Rev. SYDNEY SMITH'S MISCELLANEOUS WORKS; including
his Contributions to the *Edinburgh Review.* Crown 8vo. 6s.

The Wit and Wisdom of the Rev. Sydney Smith; a Selection of
the most memorable Passages in his Writings and Conversation. 16mo. 3s. 6d.

The ECLIPSE of FAITH; or, a Visit to a Religious Sceptic. By
HENRY ROGERS. Latest Edition. Fcp. 8vo. price 5s.

Defence of the Eclipse of Faith, by its Author; a rejoinder to Dr.
Newman's *Reply.* Latest Edition. Fcp 8vo. price 3s. 6d.

CHIPS from a GERMAN WORKSHOP; Essays on the Science of
Religion, and on Mythology, Traditions, and Customs. By F. MAX MÜLLER,
M.A. &c. Second Edition. 3 vols. 8vo. £2.

ANALYSIS of the PHENOMENA of the HUMAN MIND. By
JAMES MILL. A New Edition, with Notes, Illustrative and Critical, by
ALEXANDER BAIN, ANDREW FINDLATER, and GEORGE GROTE. Edited, with
additional Notes, by JOHN STUART MILL. 2 vols. 8vo. price 28s.

An INTRODUCTION to MENTAL PHILOSOPHY, on the Inductive
Method. By J. D. MORELL, M.A. LL.D. 8vo. 12s.

ELEMENTS of PSYCHOLOGY, containing the Analysis of the
Intellectual Powers. By J. D. MORELL, M.A. LL.D. Post 8vo. 7s. 6d.

The SECRET of HEGEL; being the Hegelian System in Origin,
Principle, Form, and Matter. By J. H. STIRLING, LL.D. 2 vols. 8vo. 28s.

SIR WILLIAM HAMILTON; being the Philosophy of Perception: an
Analysis. By J. H. STIRLING, LL.D. 8vo. 5s.

The SENSES and the INTELLECT. By ALEXANDER BAIN, M.D.
Professor of Logic in the University of Aberdeen. Third Edition. 8vo. 15s.

MENTAL and MORAL SCIENCE: a Compendium of Psychology
and Ethics. By the same Author. Third Edition. Crown 8vo. 10s. 6d. Or
separately : PART I. *Mental Science,* 6s. 6d. PART II. *Moral Science,* 4s. 6d.

LOGIC, DEDUCTIVE and INDUCTIVE. By the same Author. In
TWO PARTS, crown 8vo. 10s. 6d. Each Part may be had separately :—
PART I. *Deduction,* 4s. PART II. *Induction,* 6s. 6d.

TIME and SPACE; a Metaphysical Essay. By SHADWORTH H.
HODGSON. (This work covers the whole ground of Speculative Philosophy.)
8vo. price 16s.

The THEORY of PRACTICE; an ETHICAL ENQUIRY. By the same
Author. (This work, in conjunction with the foregoing, completes a system of
Philosophy.) 2 vols. 8vo. price 24s.

The PHILOSOPHY of NECESSITY; or, Natural Law as applicable to
Mental, Moral, and Social Science. By CHARLES BRAY. 8vo. 9s.

On Force, its Mental and Moral Correlates. By the same Author.
8vo. 5s.

A MANUAL of ANTHROPOLOGY, or **SCIENCE of MAN,** based on Modern Research. By CHARLES BRAY. Crown 8vo. price 6s.

A PHRENOLOGIST AMONGST the TODAS, or the Study of a Primitive Tribe in South India; History, Character, Customs, Religion, Infanticide, Polyandry, Language. By W. E. MARSHALL, Lieutenant-Colonel B.S.C. With 26 Illustrations. 8vo 21s.

A TREATISE on HUMAN NATURE; being an Attempt to Introduce the Experimental Method of Reasoning into Moral Subjects. By DAVID HUME. Edited, with Notes, &c. by T. H. GREEN, Fellow of Balliol College, Oxford; and T. H. GROSE, Fellow and Tutor of Queen's College, Oxford. 2 vols. 8vo.
[*In the press.*

ESSAYS MORAL, POLITICAL, and LITERARY. By DAVID HUME. By the same Editors. 2 vols. 8vo. [*In the press.*

UEBERWEG'S SYSTEM of LOGIC and HISTORY of LOGICAL DOCTRINES. Translated, with Notes and Appendices, by T. M. LINDSAY, M.A. F.R.S.E. 8vo. price 16s.

A BUDGET of PARADOXES. By AUGUSTUS DE MORGAN, F.R.A.S. and C.P.S. 8vo. 15s.

The O'KEEFFE CASE; a full Report of the Case of the Rev. Robert O'Keeffe v. Cardinal Cullen, including the Evidence and the Judgments. With an Introduction by H. C. KIRKPATRICK, Barrister. 8vo. 12s.

Astronomy, Meteorology, Popular Geography, &c.

BRINKLEY'S ASTRONOMY. Revised and partly re-written, with Additional Chapters, and an Appendix of Questions for Examination. By J. W. STUBBS, D.D. Fellow and Tutor of Trinity College, Dublin, and F. BRUNNOW, Ph.D. Astronomer Royal of Ireland. Crown 8vo. price 6s.

OUTLINES of ASTRONOMY. By Sir J. F. W. HERSCHEL, Bart. M.A. Latest Edition, with Plates and Diagrams. Square crown 8vo. 12s.

ESSAYS on ASTRONOMY: a Series of Papers on Planets and Meteors, the Sun and Sun-surrounding Space, Stars and Star-Cloudlets; and a Dissertation on the approaching Transit of Venus. By RICHARD A. PROCTOR, B.A. With 10 Plates and 24 Woodcuts. 8vo. 12s.

The UNIVERSE and the COMING TRANSITS: Presenting Researches into and New Views respecting the Constitution of the Heavens; together with an Investigation of the Conditions of the Coming Transits of Venus. By R. A. PROCTOR, B.A. With 22 Charts and 22 Woodcuts. 8vo. 16s.

The MOON; her Motions, Aspect, Scenery, and Physical Condition. By R. A. PROCTOR, B.A. With Plates, Charts, Woodcuts, and Three Lunar Photographs. Crown 8vo. 15s.

The SUN; RULER, LIGHT, FIRE, and LIFE of the PLANETARY SYSTEM. By R. A. PROCTOR, B.A. Second Edition, with 10 Plates (7 coloured) and 107 Figures on Wood. Crown 8vo. 14s.

OTHER WORLDS THAN OURS; the Plurality of Worlds Studied under the Light of Recent Scientific Researches. By R. A. PROCTOR, B.A. Third Edition, with 14 Illustrations. Crown 8vo. 10s. 6d.

The ORBS AROUND US; a Series of Familiar Essays on the Moon and Planets, Meteors and Comets, the Sun and Coloured Pairs of Stars. By R. A. PROCTOR, B.A. Crown 8vo. price 7s. 6d.

SATURN and its SYSTEM. By R. A. PROCTOR, B.A. 8vo. with 14 Plates, 14s.

SCHELLEN'S SPECTRUM ANALYSIS, in its application to Terrestrial Substances and the Physical Constitution of the Heavenly Bodies. Translated by JANE and C. LASSELL; edited, with Notes, by W. HUGGINS, LL.D. F.R.S. With 13 Plates (6 coloured) and 223 Woodcuts. 8vo. price 28s.

A NEW STAR ATLAS, for the Library, the School, and the Observatory, in Twelve Circular Maps (with Two Index Plates). Intended as a Companion to 'Webb's Celestial Objects for Common Telescopes.' With a Letterpress Introduction on the Study of the Stars, illustrated by 9 Diagrams. By R. A. PROCTOR, B.A. Crown 8vo. 5s.

CELESTIAL OBJECTS for COMMON TELESCOPES. By the Rev. T. W. WEBB, M.A. F.R.A.S. Third Edition, revised and enlarged; with Maps, Plate, and Woodcuts. Crown 8vo. price 7s. 6d.

AIR and RAIN; the Beginnings of a Chemical Climatology. By ROBERT ANGUS SMITH, Ph.D. F.R.S. F.C.S. With 8 Illustrations. 8vo. 24s.

NAUTICAL SURVEYING, an INTRODUCTION to the PRACTICAL and THEORETICAL STUDY of. By J. K. LAUGHTON, M.A. Small 8vo. 6s.

MAGNETISM and DEVIATION of the COMPASS. For the Use of Students in Navigation and Science Schools. By J. MERRIFIELD, LL.D. 18mo. 1s. 6d.

DOVE'S LAW of STORMS, considered in connexion with the Ordinary Movements of the Atmosphere. Translated by R. H. SCOTT, M.A. 8vo. 10s. 6d.

KEITH JOHNSTON'S GENERAL DICTIONARY of GEOGRAPHY, Descriptive, Physical, Statistical, and Historical; forming a complete Gazetteer of the World. New Edition, revised and corrected to the Present Date by the Author's Son, KEITH JOHNSTON, F.R.G.S. 1 vol. 8vo.   [Nearly ready.

The POST OFFICE GAZETTEER of the UNITED KINGDOM. Being a Complete Dictionary of all Cities, Towns, Villages, and of the Principal Gentlemen's Seats, in Great Britain and Ireland; Referred to the nearest Post Town, Railway and Telegraph Station : with Natural Features and Objects of Note. By J. A. SHARP. 1 vol. 8vo. of about 1,500 pages.   [In the press.

The PUBLIC SCHOOLS ATLAS of MODERN GEOGRAPHY. In 31 Maps, exhibiting clearly the more important Physical Features of the Countries delineated, and Noting all the Chief Places of Historical, Commercial, or Social Interest. Edited, with an Introduction, by the Rev. G. BUTLER, M.A. Imp. 4to. price 3s. 6d. sewed, or 5s. cloth.

The PUBLIC SCHOOLS MANUAL of MODERN GEOGRAPHY. By the Rev. GEORGE BUTLER, M.A. Principal of Liverpool College; Editor of 'The Public Schools Atlas of Modern Geography.'   [In preparation.

The PUBLIC SCHOOLS ATLAS of ANCIENT GEOGRAPHY Edited, with an Introduction on the Study of Ancient Geography, by the Rev. GEORGE BUTLER, M.A. Principal of Liverpool College. Imperial Quarto.
          [In preparation.

A MANUAL of GEOGRAPHY, Physical, Industrial, and Political. By W. HUGHES, F.R.G.S. With 6 Maps. Fcp. 7s. 6d.

MAUNDER'S TREASURY of GEOGRAPHY, Physical, Historical, Descriptive, and Political. Edited by W. HUGHES, F.R.G.S. Revised Edition, with 7 Maps and 16 Plates. Fcp. 6s. cloth, or 10s. bound in calf.

Natural History and Popular Science.

LONGMAN & CO.'S TEXT-BOOKS of SCIENCE, MECHANICAL and
PHYSICAL, adapted for the use of Artisans and of Students in Public and
Science Schools:—
ANDERSON's Strength of Materials, small 8vo. 3s. 6d.
ARMSTRONG's Organic Chemistry, 3s. 6d.
BLOXAM's Metals, 3s. 6d.
GOODEVE's Elements of Mechanism, 3s. 6d.
———— Principles of Mechanics, 3s. 6d.
GRIFFIN's Algebra and Trigonometry, 3s. 6d. Notes, 3s.6d.
JENKIN's Electricity and Magnetism, 3s. 6d.
MAXWELL's Theory of Heat, 3s. 6d.
MERRIFIELD's Technical Arithmetic and Mensuration, 3s. 6d. Key, 3s. 6d.
MILLER's Inorganic Chemistry, 3s. 6d.
SHELLEY's Workshop Appliances, 3s. 6d.
THORPE's Quantitative Chemical Analysis, 4s. 6d.
THORPE & MUIR's Qualitative Analysis, 3s. 6d.
WATSON's Plane and Solid Geometry, 3s. 6d.
₊ Other Text-Books in active preparation.

ELEMENTARY TREATISE on PHYSICS, Experimental and Applied.
Translated and edited from GANOT's Éléments de Physique by E. ATKINSON,
Ph.D. F.C.S. New Edition, revised and enlarged; with a Coloured Plate and
726 Woodcuts. Post 8vo. 15s.

NATURAL PHILOSOPHY for GENERAL READERS and YOUNG
PERSONS; being a Course of Physics divested of Mathematical Formulæ
expressed in the language of daily life. Translated from GANOT's Cours de
Physique and by E. ATKINSON, Ph.D. F.C.S. Crown 8vo. with 404 Woodcuts,
price 7s. 6d.

HELMHOLTZ'S POPULAR LECTURES on SCIENTIFIC SUBJECTS.
Translated by E. ATKINSON, Ph.D. F.C.S. Professor of Experimental Science,
Staff College. With an Introduction by Professor TYNDALL. 8vo. with nume-
rous Woodcuts, price 12s. 6d.

SOUND: a Course of Eight Lectures delivered at the Royal Institution
of Great Britain. By JOHN TYNDALL, LL.D. D.C.L. F.R.S. New Edition,
with 169 Woodcuts. Crown 8vo. 9s.

HEAT a MODE of MOTION. By JOHN TYNDALL, LL.D. D.C.L.
F.R.S. Fourth Edition. Crown 8vo. with Woodcuts, 10s. 6d.

CONTRIBUTIONS to MOLECULAR PHYSICS in the DOMAIN of
RADIANT HEAT. By J. TYNDALL, LL.D. D.C.L. F.R.S. With 2 Plates and
31 Woodcuts. 8vo. 16s.

RESEARCHES on DIAMAGNETISM and MAGNE-CRYSTALLIC
ACTION; including the Question of Diamagnetic Polarity. By J. TYNDALL,
M.D. D.C.L. F.R.S. With 6 plates and many Woodcuts. 8vo. 14s.

NOTES of a COURSE of SEVEN LECTURES on ELECTRICAL
PHENOMENA and THEORIES. delivered at the Royal Institution, A.D. 1870.
By JOHN TYNDALL, LL.D., D.C.L., F.R.S. Crown 8vo. 1s. sewed; 1s. 6d. cloth.

ELEMENTARY TREATISE on the WAVE-THEORY of LIGHT.
By HUMPHREY LLOYD, D.D. D.C.L. Provost of Trinity College, Dublin. Third
Edition, revised and enlarged. 8vo. price 10s. 6d.

LECTURES on LIGHT delivered in the United States of America in
the Years 1872 and 1873. By JOHN TYNDALL, LL.D. D.C.L. F.R.S. With
Frontispiece and Diagrams. Crown 8vo. price 7s. 6d.

NOTES of a COURSE of NINE LECTURES on LIGHT delivered at the
Royal Institution, A.D. 1869.   By JOHN TYNDALL, LL.D. D.C.L. F.R.S.
Crown 8vo. price 1s. sewed, or 1s. 6d. cloth.

FRAGMENTS of SCIENCE.   By JOHN TYNDALL, LL.D. D.C.L. F.R.S.
Third Edition.   8vo. price 14s.

LIGHT SCIENCE for LEISURE HOURS; a Series of Familiar
Essays on Scientific Subjects, Natural Phenomena, &c.   By R. A. PROCTOR,
B.A.   First and Second Series.   Crown 8vo. 7s. 6d. each.

The CORRELATION of PHYSICAL FORCES.   By the Hon. Sir W. R.
GROVE, M.A. F.R.S. &c.   Sixth Edition, with other Contributions to Science.
8vo.   [In the press.

Professor OWEN'S LECTURES on the COMPARATIVE ANATOMY
and Physiology of the Invertebrate Animals.   Second Edition, with 235 Woodcuts.
8vo. 21s.

The COMPARATIVE ANATOMY and PHYSIOLOGY of the VERTE-
BRATE ANIMALS.   By RICHARD OWEN, F.R.S. D.C.L.   With 1,472 Woodcuts.
3 vols. 8vo. £3. 13s. 6d.

PRINCIPLES of ANIMAL MECHANICS.   By the Rev. S. HAUGHTON,
F.R.S. Fellow of Trin. Coll. Dubl. M.D. Dubl. and D.C.L. Oxon.   Second
Edition, with 111 Figures on Wood.   8vo. 21s.

The EARTH and MAN; or, Physical Geography in relation to the
History of Mankind.   Slightly Abridged from the French of A. GUIZOT, with a
few Notes.   Fifth Edition.   Fcp. 8vo. 2s.

ROCKS CLASSIFIED and DESCRIBED.   By BERNHARD VON COTTA.
English Edition, by P. H. LAWRENCE; with English, German, and French
Synonymes.   Post 8vo. 14s.

GEOLOGY SIMPLIFIED for BEGINNERS.   By A. C. RAMSAY, LL.D.
F.R.S.   Forming part of the Rev. G. R. Gleig's New School Series.   18mo.
                                                          [In the press.

The ANCIENT STONE IMPLEMENTS, WEAPONS, and ORNA-
MENTS of GREAT BRITAIN.   By JOHN EVANS, F.R.S. F.S.A.   With 2 Plates
and 476 Woodcuts.   8vo. price 28s.

The ORIGIN of CIVILISATION and the PRIMITIVE CONDITION
of MAN; Mental and Social Condition of Savages.   By Sir JOHN LUBBOCK,
Bart. M.P. F.R.S.   Second Edition, with 25 Woodcuts.   8vo. price 16s.

BIBLE ANIMALS; being a Description of every Living Creature
mentioned in the Scriptures, from the Ape to the Coral.   By the Rev. J. G.
WOOD, M.A. F.L.S.   With about 100 Vignettes on Wood.   8vo. 21s.

HOMES WITHOUT HANDS; a Description of the Habitations of
Animals, classed according to their Principle of Construction.   By the Rev. J.
G. WOOD, M.A. F.L.S.   With about 140 Vignettes on Wood.   8vo. 21s.

INSECTS AT HOME; a Popular Account of British Insects, their
Structure, Habits, and Transformations.   By the Rev. J. G. WOOD, M.A. F.L.S.
With upwards of 700 Illustrations.   8vo. price 21s.

INSECTS ABROAD; a Popular Account of Foreign Insects, their
Structure, Habits, and Transformations.   By J. G. WOOD, M.A. F.L.S. Printed
and illustrated uniformly with 'Insects at Home,' to which it will form a
Sequel and Companion.   [In the press.

STRANGE DWELLINGS; a description of the Habitations of Animals, abridged from 'Homes without Hands.' By the Rev. J. G. WOOD, M.A. F.L.S. With about 60 Woodcut Illustrations. Crown 8vo. price 7s. 6d.

OUT of DOORS; a Series of Essays on Natural History. By the Rev. J. G. WOOD, M.A. F.L.S. With Six Illustrations from Original Designs engraved on Wood by G. Pearson. Crown 8vo. [Nearly ready.

A FAMILIAR HISTORY of BIRDS. By E. STANLEY, D.D. F.R.S. late Lord Bishop of Norwich. Seventh Edition, with Woodcuts. Fcp. 3s. 6d.

FROM JANUARY to DECEMBER; a Book for Children. Second Edition. 8vo. 3s. 6d.

The HARMONIES of NATURE and UNITY of CREATION. By Dr. GEORGE HARTWIG. 8vo. with numerous Illustrations, 18s.

The SEA and its LIVING WONDERS. By Dr. GEORGE HARTWIG. Latest revised Edition. 8vo. with many Illustrations, 10s. 6d.

The TROPICAL WORLD. By Dr. GEORGE HARTWIG. With above 160 Illustrations. Latest revised Edition. 8vo. price 10s. 6d.

The SUBTERRANEAN WORLD. By Dr. GEORGE HARTWIG. With 3 Maps and about 80 Woodcuts, including 8 full size of page. 8vo. price 21s.

The POLAR WORLD, a Popular Description of Man and Nature in the Arctic and Antarctic Regions of the Globe. By Dr. GEORGE HARTWIG. With 8 Chromoxylographs, 3 Maps, and 85 Woodcuts. 8vo. 10s. 6d.

KIRBY and SPENCE'S INTRODUCTION to ENTOMOLOGY, or Elements of the Natural History of Insects. 7th Edition. Crown 8vo. 5s.

MAUNDER'S TREASURY of NATURAL HISTORY, or Popular Dictionary of Birds, Beasts, Fishes, Reptiles, Insects, and Creeping Things. With above 900 Woodcuts. Fcp. 8vo. price 6s. cloth, or 10s. bound in calf.

HANDBOOK of HARDY TREES, SHRUBS, and HERBACEOUS PLANTS, containing Descriptions, Native Countries, &c. of a Selection of the Best Species in Cultivation; together with Cultural Details, Comparative Hardiness, Suitability for Particular Positions, &c. By W. B. HEMSLEY, formerly Assistant at the Herbarium of the Royal Gardens, Kew. Based on DECAISNE and NAUDIN'S Manuel de l'Amateur des Jardins, and including the 264 Original Woodcuts. Medium 8vo. 21s.

A GENERAL SYSTEM of BOTANY DESCRIPTIVE and ANALYTICAL. I. Outlines of Organography, Anatomy, and Physiology; II. Descriptions and Illustrations of the Orders. By E. LE MAOUT, and J. DECAISNE, Members of the Institute of France. Translated by Mrs. HOOKER. The Orders arranged after the Method followed in the Universities and Schools of Great Britain, its Colonies, America, and India; with an Appendix on the Natural Method, and other Additions, by J. D. HOOKER, F.R.S. &c. Director of the Royal Botanical Gardens, Kew. With 5,500 Woodcuts. Imperial 8vo. price 52s. 6d.

The TREASURY of BOTANY, or Popular Dictionary of the Vegetable Kingdom; including a Glossary of Botanical Terms. Edited by J. LINDLEY, F.R.S. and T. MOORE, F.L.S. assisted by eminent Contributors. With 274 Woodcuts and 20 Steel Plates. Two Parts, fcp. 8vo. 12s. cloth, or 20s. calf.

The ELEMENTS of BOTANY for FAMILIES and SCHOOLS. Tenth Edition, revised by THOMAS MOORE, F.L.S. Fcp. with 154 Woodcuts, 2s. 6d.

The ROSE AMATEUR'S GUIDE. By THOMAS RIVERS. Fourteenth Edition. Fcp. 8vo. 4s.

LOUDON'S ENCYCLOPÆDIA of PLANTS ; comprising the Specific Character, Description, Culture, History, &c. of all the Plants found in Great Britain. With upwards of 12,000 Woodcuts. 8vo. 42s.

MAUNDER'S SCIENTIFIC and LITERARY TREASURY. New Edition. thoroughly revised and in great part rewritten, with above 1,000 new Articles, by J. Y. JOHNSON, Corr. M.Z.S. Fcp. 6s. cloth, or 10s. calf.

A DICTIONARY of SCIENCE, LITERATURE, and ART. Fourth Edition. re-edited by W. T. BRANDE (the original Author). and GEORGE W. COX, M.A., assisted by contributors of eminent Scientific and Literary Acquirements. 3 vols. medium 8vo. price 63s. cloth.

Chemistry, Medicine, Surgery, and the Allied Sciences.

A DICTIONARY of CHEMISTRY and the Allied Branches of other Sciences. By HENRY WATTS, F.R.S. assisted by eminent Contributors. Complete in 6 vols. medium 8vo. price £8. 14s. 6d. SUPPLEMENT *in the Press.*

ELEMENTS of CHEMISTRY, Theoretical and Practical. By W. ALLEN MILLER, M.D. late Prof. of Chemistry, King's Coll. London. New Edition. 3 vols. 8vo. £3. PART I. CHEMICAL PHYSICS, 15s. PART II. INORGANIC CHEMISTRY, 21s. PART III. ORGANIC CHEMISTRY, 24s.

A Course of Practical Chemistry, for the use of Medical Students. By W. ODLING, F.R.S. New Edition, with 70 Woodcuts. Crown 8vo. 7s. 6d.

A MANUAL of CHEMICAL PHYSIOLOGY, including its Points of Contact with Pathology. By J. L. W. THUDICHUM, M.D. With Woodcuts. 8vo. price 7s. 6d.

SELECT METHODS in CHEMICAL ANALYSIS, chiefly INORGANIC. By WILLIAM CROOKES, F.R.S. With 22 Woodcuts. Crown 8vo. price 12s. 6d.

A HANDBOOK of DYEING and CALICO PRINTING. By WILLIAM CROOKES, F.R.S. Illustrated with numerous Specimens of Dyed Textile Fabrics. 8vo. [*In the Spring.*

LECTURES on the DISEASES of INFANCY and CHILDHOOD. By CHARLES WEST, M.D. &c. Sixth Edition, revised and enlarged. 8vo. 18s.

The SCIENCE and ART of SURGERY ; being a Treatise on Surgical Injuries, Diseases. and Operations. By JOHN ERIC ERICHSEN, Senior Surgeon to University College Hospital, and Holme Professor of Clinical Surgery in University College, London. The Sixth Edition, with 712 Woodcuts. 2 vols. 8vo. price 32s.

A SYSTEM of SURGERY, Theoretical and Practical. In Treatises by Various Authors. Edited by T. HOLMES, M.A. &c. Surgeon and Lecturer on Surgery at St. George's Hospital. Second Edition, thoroughly revised, with numerous Illustrations. 5 vols. 8vo. £5. 5s.

The SURGICAL TREATMENT of CHILDREN'S DISEASES. By T. HOLMES, M.A. &c. late Surgeon to the Hospital for Sick Children. Second Edition, with 9 plates and 112 Woodcuts. 8vo. 21s.

LECTURES on the PRINCIPLES and PRACTICE of PHYSIC. By
Sir THOMAS WATSON, Bart. M.D.   Fifth Edition, thoroughly revised.
2 vols. 8vo. price 36s.

LECTURES on SURGICAL PATHOLOGY. By Sir JAMES PAGET,
Bart. F.R.S.   Third Edition, revised and re-edited by the Author and
Professor W. TURNER, M.D.   8vo. with 134 Woodcuts, 21s.

On the SURGICAL DISEASES of the TEETH and CONTIGUOUS
STRUCTURES, with their Treatment.   By S. JAMES A. SALTER, M.B. F.R.S.
Dental Surgeon to Guy's Hospital.   8vo. with numerous Illustrations.
[*In the Autumn.*

A TREATISE on MEDICAL ELECTRICITY, THEORETICAL and
Practical; and its Use in the Treatment of Paralysis, Neuralgia, and other
Diseases.   By JULIUS ALTHAUS, M.D. M.R.C.P. &c.   Third Edition, enlarged
and revised; with 147 Illustrations. 8vo. price 18s.

LECTURES on FEVER delivered in the Theatre of the Meath Hospital
and County of Dublin Infirmary.   By W. STOKES. M.D. F.R.S. Physician to the
Queen in Ireland.   Edited by J. W. MOORE, M.D. F.K.Q.C.P.   8vo. 15s.

The SKIM-MILK TREATMENT of DIABETES and BRIGHT'S
DISEASE; with Clinical Observations on the Symptoms and Pathology of
these Affections.   By A. S. DONKIN, M.D. &c.   Crown 8vo. 10s. 6d.

QUAIN'S ELEMENTS of ANATOMY.   Seventh Edition [1867],
edited by W. SHARPEY, M.D. F.R.S.   ALLEN THOMSON, M.D. F.R.S. and
J. CLELAND, M.D.   With upwards of 800 Engravings on Wood.   2 vols. 8vo.
price 31s. 6d.

ANATOMY, DESCRIPTIVE and SURGICAL.   By HENRY GRAY,
F.R.S.   With about 400 Woodcuts from Dissections.   Sixth Edition, by
T. HOLMES, M.A., with a new Introduction by the Editor.   Royal 8vo. 28s.

A TREATISE on the CONTINUED FEVERS of GREAT BRITAIN.
By CHARLES MURCHISON, M.D. LL.D. F.R.S. F.R.C.P. &c.   Second Edition,
revised and enlarged, with numerous Illustrations.   8vo. price 24s.

CLINICAL LECTURES on DISEASES of the LIVER, JAUNDICE,
and ABDOMINAL DROPSY.   By CHARLES MURCHISON, M.D. &c.   New
Edition, preparing for publication.

OUTLINES of PHYSIOLOGY, Human and Comparative.   By JOHN
MARSHALL, F.R.C.S. Surgeon to the University College Hospital.   2 vols.
crown 8vo. with 122 Woodcuts, 32s.

PHYSIOLOGICAL ANATOMY and PHYSIOLOGY of MAN.   By the
late R. B. TODD, M.D. F.R.S. and W. BOWMAN, F.R.S. of King's College.
With numerous Illustrations.   Vol. II. 8vo. 25s.

VOL. I. New Edition by Dr. LIONEL S. BEALE, F.R.S. in course of publication, with many Illustrations.   PARTS I. and II. price 7s. 6d. each.

COPLAND'S DICTIONARY of PRACTICAL MEDICINE, abridged
from the larger work and throughout brought down to the present State
of Medical Science.   8vo. 36s.

DR. PEREIRA'S ELEMENTS of MATERIA MEDICA and THERA-
PEUTICS, abridged and adapted for the use of Medical and Pharmaceutical
Practitioners and Students; and comprising all the Medicines of the
British Pharmacopœia, with such others as are frequently ordered in Pre-
scriptions or required by the Physician.   Edited by Professor BENTLEY,
F.L.S. &c. and by Dr. REDWOOD, F.C.S. &c.   With 125 Woodcut Illustra-
tions.   8vo. price 25s.

B

The ESSENTIALS of MATERIA MEDICA and THERAPEUTICS.
By ALFRED BARING GARROD, M.D. F.R.S. &c. Physician to King's College
Hospital. Third Edition. Seventh Impression, brought up to 1870. Crown
8vo. price 12s. 6d.

The Fine Arts, and Illustrated Editions.

A DICTIONARY of ARTISTS of the ENGLISH SCHOOL: Painters,
Sculptors, Architects, Engravers, and Ornamentists; with Notices of their Lives
and Works. By S. REDGRAVE. 8vo. 16s.

The THREE CATHEDRALS DEDICATED to ST. PAUL, in LONDON;
their History from the Foundation of the First Building in the Sixth Century
to the Proposals for the Adornment of the Present Cathedral. By WILLIAM
LONGMAN, F.A.S. with numerous Illustrations. Square crown 8vo. price 21s.

GROTESQUE ANIMALS, invented, described, and portrayed by E. W.
COOKE, R.A. F.R.S. F.G.S. F.Z.S. &c. in Twenty-four Plates, with Elucidatory
Comments. Royal 4to. 21s.

IN FAIRYLAND; Pictures from the Elf-World. By RICHARD
DOYLE. With a Poem by W. ALLINGHAM. With Sixteen Plates, containing
Thirty-six Designs printed in Colours. Folio, 31s. 6d.

ALBERT DURER, HIS LIFE and WORKS; including Auto-
biographical Papers and Complete Catalogues. By WILLIAM B. SCOTT.
With Six Etchings by the Author, and other Illustrations. 8vo. 16s.

The NEW TESTAMENT, illustrated with Wood Engravings after the
Early Masters, chiefly of the Italian School. Crown 4to. 63s. cloth, gilt top;
or £5 5s. elegantly bound in morocco.

LYRA GERMANICA; the Christian Year. Translated by CATHERINE
WINKWORTH; with 125 Illustrations on Wood drawn by J. LEIGHTON,
F.S.A. 4to. 21s.

LYRA GERMANICA; the Christian Life. Translated by CATHERINE
WINKWORTH; with about 200 Woodcut Illustrations by J. LEIGHTON, F.S.A.
and other Artists. 4to. 21s.

The LIFE of MAN SYMBOLISED by the MONTHS of the YEAR.
Text selected by R. PIGOT; Illustrations on Wood from Original Designs by
J. LEIGHTON, F.S.A. 4to. 42s.

CATS' and FARLIE'S MORAL EMBLEMS; with Aphorisms, Adages,
and Proverbs of all Nations. 121 Illustrations on Wood by J. LEIGHTON,
F.S.A. Text selected by R. PIGOT. Imperial 8vo. 31s. 6d.

SACRED and LEGENDARY ART. By MRS. JAMESON.

Legends of the Saints and Martyrs. New Edition, with 19
Etchings and 187 Woodcuts. 2 vols. square crown 8vo. 31s. 6d.

Legends of the Monastic Orders. New Edition, with 11 Etchings
and 88 Woodcuts. 1 vol. square crown 8vo. 21s.

Legends of the Madonna. New Edition, with 27 Etchings and
165 Woodcuts. 1 vol. square crown 8vo. 21s.

The History of Our Lord, with that of his Types and Precursors.
Completed by Lady EASTLAKE. Revised Edition, with 31 Etchings and
281 Woodcuts. 2 vols. square crown 8vo. 42s.

The Useful Arts, Manufactures, &c.

HISTORY of the GOTHIC REVIVAL; an Attempt to shew how far the taste for Mediæval Architecture was retained in England during the last-two centuries, and has been re-developed in the present. By C. L. EAST-LAKE, Architect. With 48 Illustrations Imperial 8vo. 31s. 6d.

GWILT'S ENCYCLOPÆDIA of ARCHITECTURE, with above 1,600 Engravings on Wood. Fifth Edition, revised and enlarged by WYATT PAPWORTH. 8vo. 52s. 6d.

A MANUAL of ARCHITECTURE: being a Concise History and Explanation of the principal Styles of European Architecture. Ancient, Mediæval, and Renaissance; with a Glossary of Technical Terms. By THOMAS MITCHELL. Crown 8vo. with 150 Woodcuts, 10s. 6d.

HINTS on HOUSEHOLD TASTE in FURNITURE, UPHOLSTERY, and other Details. By CHARLES L. EASTLAKE, Architect. New Edition, with about 90 Illustrations. Square crown 8vo. 14s.

PRINCIPLES of MECHANISM, designed for the Use of Students in the Universities, and for Engineering Students generally. By R. WILLIS, M.A. F.R.S. &c. Jacksonian Professor in the University of Cambridge. Second Edition, enlarged; with 374 Woodcuts. 8vo. 18s.

GEOMETRIC TURNING: comprising a Description of Plant's New Geometric Chuck, with directions for its use, and a series of Patterns cut by it, with Explanations. By H. S. SAVORY. With numerous Woodcuts. 8vo. 21s.

LATHES and TURNING, Simple, Mechanical, and ORNAMENTAL. By W. HENRY NORTHCOTT. With about 240 Illustrations on Steel and Wood. 8vo. 18s.

PERSPECTIVE; or, the Art of Drawing what One Sees. Explained and adapted to the use of those Sketching from Nature.. By Lieut. W. H. COLLINS, R.E. F.R.A.S. With 37 Woodcuts. Crown 8vo. price 5s.

INDUSTRIAL CHEMISTRY; a Manual for Manufacturers and for use in Colleges or Technical Schools. Being a Translation of Professors Stohmann and Engler's German Edition of PAYEN's Précis de Chimie Industrielle, by Dr. J. D. BARRY. Edited and supplemented by B. H. PAUL, Ph.D. 8vo. with Plates and Woodcuts. [In the press.

URE'S DICTIONARY of ARTS, MANUFACTURES, and MINES. Sixth Edition, rewritten and enlarged by ROBERT HUNT, F.R.S. assisted by numerous Contributors eminent in Science and the Arts, and familiar with Manufactures. With above 2,000 Woodcuts. 3 vols. medium 8vo. £4 14s. 6d.

HANDBOOK of PRACTICAL TELEGRAPHY. By R. S. CULLEY,. Memb. Inst. C.E. Engineer-in-Chief of Telegraphs to the Post Office. Sixth Edition, with 144 Woodcuts and 5 Plates. 8vo. price 16s.

The ENGINEER'S HANDBOOK; explaining the Principles which should guide the Young Engineer in the Construction of Machinery, with the necessary Rules, Proportions, and Tables. By C. S. LOWNDES. Post 8vo. 5s.

ENCYCLOPÆDIA of CIVIL ENGINEERING, Historical, Theoretical, and Practical. By E. CRESY, C.E. With above 3,000 Woodcuts. 8vo. 42s.

The STRAINS IN TRUSSES Computed by means of Diagrams; with 20 Examples drawn to Scale. By F. A. RANKEN, M.A. C.E. With 35 Diagrams. Square crown 8vo. 6s. 6d.

TREATISE on MILLS and MILLWORK. By Sir W. FAIRBAIRN, Bart. F.R.S. New Edition, with 18 Plates and 322 Woodcuts, 2 vols. 8vo. 32s.

USEFUL INFORMATION for ENGINEERS. By Sir W. FAIRBAIRN, Bart. F.R.S. Revised Edition, with numerous Illustrations. 3 vols. crown 8vo. price 31s. 6d.

The APPLICATION of CAST and WROUGHT IRON to Building Purposes. By Sir W. FAIRBAIRN, Bart. F.R.S. Fourth Edition, enlarged; with 6 Plates and 118 Woodcuts. 8vo. price 16s.

GUNS and STEEL; Miscellaneous Papers on Mechanical Subjects. By Sir JOSEPH WHITWORTH, Bart. C.E. F.R.S. LL.D. D.C.L. Royal 8vo. with Illustrations, 7s. 6d.

A TREATISE on the STEAM ENGINE, in its various Applications to Mines, Mills, Steam Navigation, Railways, and Agriculture. By J. BOURNE, C.E. Eighth Edition; with Portrait, 37 Plates, and 546 Woodcuts. 4to. 42s.

CATECHISM of the STEAM ENGINE, in its various Applications to Mines, Mills, Steam Navigation, Railways, and Agriculture. By the same Author. With 89 Woodcuts. Fcp. 6s.

HANDBOOK of the STEAM ENGINE. By the same Author, forming a KEY to the Catechism of the Steam Engine, with 67 Woodcuts. Fcp. 9s.

BOURNE'S RECENT IMPROVEMENTS in the STEAM ENGINE in its various applications to Mines, Mills, Steam Navigation, Railways, and Agriculture. By JOHN BOURNE, C.E. New Edition including many New Examples with 124 Woodcuts. Fcp. 8vo. 6s.

PRACTICAL TREATISE on METALLURGY, adapted from the last German Edition of Professor KERL'S *Metallurgy* by W. CROOKES, F.R.S. &c. and E. BÖHRIG, Ph.D. M.E. With 625 Woodcuts. 3 vols. 8vo. price £4 19s.

MITCHELL'S MANUAL of PRACTICAL ASSAYING. Fourth Edition, for the most part rewritten, with all the recent Discoveries incorporated, by W. CROOKES, F.R.S. With 199 Woodcuts. 8vo. 31s. 6d.

LOUDON'S ENCYCLOPÆDIA of AGRICULTURE: comprising the Laying-out, Improvement, and Management of Landed Property, and the Cultivation and Economy of Agricultural Produce. With 1,100 Woodcuts. 8vo. 21s.

Loudon's Encyclopædia of Gardening: comprising the Theory and Practice of Horticulture, Floriculture, Arboriculture, and Landscape Gardening. With 1,000 Woodcuts. 8vo. 21s.

BAYLDON'S ART of VALUING RENTS and TILLAGES, and Claims of Tenants upon Quitting Farms, both at Michaelmas and Lady Day. Eighth Edition, revised by J. C. MORTON. 8vo. 10s. 6d.

Religious and *Moral* *Works.*

INTRODUCTION to the SCIENCE of RELIGION. Four Lectures delivered at the Royal Institution; with Two Essays on False Analogies and the Philosophy of Mythology. By F. MAX MÜLLER, M.A. Crown 8vo. 10s. 6d.

SUPERNATURAL RELIGION; an Inquiry into the Reality of Divine Revelation. 2 vols. 8vo. 24s.

ESSAYS on the HISTORY of the CHRISTIAN RELIGION. By JOHN Earl RUSSELL. Cabinet Edition, revised. Fcp. 8vo. price 3s. 6d.

The **SPEAKER'S BIBLE COMMENTARY**, by Bishops and other
Clergy of the Anglican Church, critically examined by the Right Rev. J. W.
COLENSO, D.D. Bishop of Natal. 8vo. PART I. *Genesis*, 3*s*. 6*d*. PART II.
Exodus, 4*s*. 6*d*. PART III. *Leviticus*, 2*s*. 6*d*. PART IV. *Numbers*, 3*s*. 6*d*.
PART V. *Deuteronomy*, 5*s*.

The **OUTLINES of the CHRISTIAN MINISTRY DELINEATED**, and
brought to the Test of Reason, Holy Scripture History, and Experience, with a
view to the Reconciliation of Existing Differences concerning it, especially
between Presbyterians and Episcopalians. By C. WORDSWORTH, D.C.L. Bishop
of St. Andrews. Crown 8vo. price 7*s*. 6*d*.

CHRIST the CONSOLER; a Book of Comfort for the Sick. With a
Preface by the Right Rev. the Lord Bishop of Carlisle. Small 8vo. price 6*s*.

REASONS of FAITH; or, the ORDER of the Christian Argument
Developed and Explained. By the Rev. G. S. DREW, M.A. Second Edition,
revised and enlarged. Fcp. 8vo. price 6*s*.

SYNONYMS of the OLD TESTAMENT, their BEARING on CHRIS-
TIAN FAITH and PRACTICE. By the Rev. R. B. GIRDLESTONE, M.A. 8vo. 15*s*.

The **ANTIQUITIES of ISRAEL**. By HEINRICH EWALD, Professor of
the University of Göttingen. Translated from the German. 8vo. [*In the press.*

An **INTRODUCTION to the THEOLOGY of the CHURCH of**
ENGLAND, in an Exposition of the Thirty-nine Articles. By the Rev. T. P.
BOULTBEE, LL.D. New Edition, Fcp. 8vo. price 6*s*.

FUNDAMENTALS; or, Bases of Belief concerning MAN and GOD :
a Handbook of Mental, Moral, and Religious Philosophy. By the Rev. T.
GRIFFITH, M.A. 8vo. price 10*s*. 6*d*.

SERMONS for the TIMES preached in St. Paul's Cathedral and
elsewhere. By the Rev. THOMAS GRIFFITH, M.A. Prebendary of St. Paul's.
Crown 8vo. 6*s*.

PRAYERS for the FAMILY and for PRIVATE USE, selected
from the COLLECTION of the late BARON BUNSEN, and Translated by
CATHERINE WINKWORTH. Fcp. 8vo. price 3*s*. 6*d*.

An **EXPOSITION of the 39 ARTICLES**, Historical and Doctrinal.
By E. HAROLD BROWNE, D.D. Lord Bishop of Winchester. Ninth Edit. 8vo. 16*s*.

The **LIFE and EPISTLES of ST. PAUL**. By the Rev. W. J.
CONYBEARE, M.A., and the Very Rev. J. S. HOWSON, D.D. Dean of Chester :—
    LIBRARY EDITION, with all the Original Illustrations, Maps, Landscapes on
    Steel, Woodcuts, &c. 2 vols. 4to. 48*s*.
    INTERMEDIATE EDITION, with a Selection of Maps, Plates, and Woodcuts.
    2 vols. square crown 8vo. 21*s*.
    STUDENT'S EDITION, revised and condensed, with 46 Illustrations and Maps.
    1 vol. crown 8vo. price 9*s*.

The **VOYAGE and SHIPWRECK of ST PAUL**; with Dissertations
on the Life and Writings of St. Luke and the Ships and Navigation of the
Ancients. By JAMES SMITH, F.R.S. Third Edition. Crown 8vo. 10*s*. 6*d*.

COMMENTARY on the EPISTLE to the ROMANS. By the Rev.
W. A. O'CONOR, B.A. Rector of St. Simon and St. Jude, Manchester. Crown
8vo. price 3*s*. 6*d*.

The **EPISTLE to the HEBREWS**; with Analytical Introduction and
Notes. By the Rev. W. A. O'CONOR, B.A. Crown 8vo. price 4*s*. 6*d*.

ST. MARK'S GOSPEL; Greek Text, with English Vocabulary. Edited
the Rev. J. T. WHITE, D.D. Oxon. 32mo. 1*s*. 6*d*.

ST. JOHN'S GOSPEL; Greek Text, with English Vocabulary. Edited by the Rev. J. T. WHITE, D.D. Oxon. 32mo. *[Just ready.*

A CRITICAL and GRAMMATICAL COMMENTARY on ST. PAUL'S Epistles. By C. J. ELLICOTT, D.D. Lord Bishop of Gloucester and Bristol. 8vo.

Galatians, Fourth Edition, 8s. 6d.

Ephesians, Fourth Edition, 8s. 6d.

Pastoral Epistles, Fourth Edition, 10s. 6d.

Philippians, Colossians, and Philemon, Third Edition, 10s. 6d.

Thessalonians, Third Edition, 7s. 6d.

HISTORICAL LECTURES on the LIFE of OUR LORD. By C. J. ELLICOTT, D.D. Bishop of Gloucester and Bristol. Fifth Edition. 8vo. 12s.

EVIDENCE of the TRUTH of the CHRISTIAN RELIGION derived from the Literal Fulfilment of Prophecy. By ALEXANDER KEITH, D.D. 37th Edition, with numerous Plates, in square 8vo. 12s. 6d.; also the 39th Edition, in post 8vo. with 5 Plates, 6s.

The HISTORY and LITERATURE of the ISRAELITES, according to the Old Testament and the Apocrypha. By C. DE ROTHSCHILD and A. DE ROTHSCHILD. Second Edition, revised. 2 vols. post 8vo. with Two Maps, price 12s. 6d. Abridged Edition, in 1 vol. fcp. 8vo. price 3s. 6d.

An INTRODUCTION to the STUDY of the NEW TESTAMENT, Critical, Exegetical, and Theological. By the Rev. S. DAVIDSON, D.D. LL.D. 2 vols. 8vo. 30s.

HISTORY of ISRAEL. By H. EWALD, Prof. of the Univ. of Göttingen. Translated by J. E. CARPENTER, M.A., with a Preface by RUSSELL MARTINEAU, M.A. 5 vols. 8vo. 63s.

The TREASURY of BIBLE KNOWLEDGE; being a Dictionary of the Books, Persons, Places, Events, and other matters of which mention is made in Holy Scripture. By Rev. J. AYRE, M.A. With Maps, 16 Plates, and numerous Woodcuts. Fcp. 8vo. price 6s. cloth, or 10s. neatly bound in calf.

LECTURES on the PENTATEUCH and the MOABITE STONE; with Appendices on the Elohistic Narrative, the Original Story of the Exodus, and the Pre-Christian Cross. By the Right Rev. J. W. COLENSO, D.D. Bishop of Natal. 8vo. 12s.

The PENTATEUCH and BOOK of JOSHUA CRITICALLY EXAMINED. By the Right Rev. J. W. COLENSO, D.D. Bishop of Natal. Crown 8vo. 6s.

AUTHORITY and CONSCIENCE; a Free Debate on the Tendency of Dogmatic Theology and on the Characteristics of Faith. Edited by CONWAY MOREL. Post 8vo. 7s. 6d.

A VIEW of the SCRIPTURE REVELATIONS CONCERNING a FUTURE STATE. By RICHARD WHATELY, D.D. late Archbishop of Dublin. Ninth Edition. Fcp. 8vo. 5s.

TEXTS and THOUGHTS for CHRISTIAN MINISTERS. By J. HARDING, D.D. late Bishop of Bombay. *[In the press.*

THOUGHTS for the AGE. By ELIZABETH M. SEWELL, Author of 'Amy Herbert,' &c. New Edition, revised. Fcp. 8vo. price 5s.

Passing Thoughts on Religion. By Miss SEWELL. Fcp. 8vo. 3s 6d.

Self-Examination before Confirmation. By Miss SEWELL. 32mo. price 1s. 6d.

Readings for a Month Preparatory to Confirmation, from Writers of the Early and English Church. By Miss SEWELL. Fcp. 4s.

Readings for Every Day in Lent, compiled from the Writings of Bishop JEREMY TAYLOR. By Miss SEWELL. Fcp. 5s.

Preparation for the Holy Communion; the Devotions chiefly from the Works of JEREMY TAYLOR. By Miss SEWELL. 32mo. 3s.

THOUGHTS for the HOLY WEEK for Young Persons. By Miss SEWELL. New Edition. Fcp. 8vo. 2s.

PRINCIPLES of EDUCATION Drawn from Nature and Revelation, and applied to Female Education in the Upper Classes. By Miss SEWELL. 2 vols. fcp. 8vo. 12s. 6d.

LYRA GERMANICA, translated from the German by Miss C. WINK-WORTH. FIRST SERIES, Hymns for the Sundays and Chief Festivals. SECOND SERIES, the Christian Life. Fcp. 8vo. price 3s. 6d. each SERIES.

SPIRITUAL SONGS for the SUNDAYS and HOLIDAYS through-out the Year. By J. S. B. MONSELL, LL.D. Fcp. 8vo. 4s. 6d.

ENDEAVOURS after the CHRISTIAN LIFE: Discourses. By the Rev. J. MARTINEAU, LL.D. Fifth Edition, carefully revised. Crown 8vo. 7s. 6d.

HYMNS of PRAISE and PRAYER, collected and edited by the Rev. J. MARTINEAU, LL.D. Crown 8vo. 4s. 6d.

WHATELY'S INTRODUCTORY LESSONS on the **CHRISTIAN** Evidences. 18mo. 6d.

BISHOP JEREMY TAYLOR'S ENTIRE WORKS. With Life by BISHOP HEBER. Revised and corrected by the Rev. C. P. EDEN. Complete in Ten Volumes, 8vo. cloth, price £5. 5s.

Travels, Voyages, &c.

MEETING the SUN; a Journey all round the World through Egypt, China, Japan, and California. By WILLIAM SIMPSON, F.R.G.S. With 48 Helio-types and Wood Engravings from Drawings by the Author. Medium 8vo. 24s.

The ATLANTIC to the PACIFIC; What to see, and How to see it. By JOHN ERASTUS LESTER, M.A. Map, Plan, Woodcuts. Crown 8vo. 6s.

SLAVE-CATCHING in the INDIAN OCEAN. By Capt. COLOMB, R.N. With a Map and Illustrations. 8vo. 21s.

UNTRODDEN PEAKS and UNFREQUENTED VALLEYS; a Mid-summer Ramble among the Dolomites. By AMELIA B. EDWARDS. With a Map and 27 Wood Engravings. Medium 8vo. 21s.

The DOLOMITE MOUNTAINS; Excursions through Tyrol, Carinthia, Carniola, and Friuli, 1861-1863. By J. GILBERT and G. C. CHURCHILL, F.R.G.S. With numerous Illustrations. Square crown 8vo. 21s.

CADORE; or, TITIAN'S COUNTRY. By JOSIAH GILBERT, one of the Authors of the ' Dolomite Mountains.' With Map, Facsimile, and 40 Illus-trations. Imperial 8vo. 31s. 6d.

HOURS of EXERCISE in the ALPS. By JOHN TYNDALL. LL.D.
D.C.L. F.R.S. Third Edit'on, with 7 Woodcuts by E. Whymper. Crown 8vo.
price 12s. 6d.

The ALPINE CLUB MAP of SWITZERLAND and the ADJACENT
COUNTRIES, on the Scale of Four Miles to the Inch ; from Schaffhausen on
the North to Milan on the South, and from the Ortler Group on the East to
Geneva on the West. Constructed under the immediate superintendence of the
ALPINE CLUB, and edited by R. C. NICHOLS, F.S.A. F.R.G.S. In Four Sheets.
                                                              [Nearly ready.

MAP of the CHAIN of MONT BLANC, from an Actual Survey in
1863–1864. By ADAMS-REILLY, F.R.G.S. M.A.C. Published under the Authority of the Alpine Club. In Chromolithography on extra stout drawing-
paper 28in. x 17in. price 10s. or mounted on canvas in a folding case, 12s. 6d.

TRAVELS in the CENTRAL CAUCASUS and BASHAN. Including
Visits to Ararat and Tabreez and Ascents of Kazbek and Elbruz. By D. W.
FRESHFIELD. Square crown 8vo. with Maps, &c. 18s.

PAU and the PYRENEES. By Count HENRY RUSSELL, Member of
the Alpine Club, &c. With 2 Maps. Fcp. 8vo. price 5s.

HOW to SEE NORWAY. By Captain J. R. CAMPBELL. With Map
and 5 Woodcuts. Fcp. 8vo. price 5s.

MY WIFE and I in QUEENSLAND; Eight Years' Experience in
the Colony, with some account of Polynesian Labour. By CHARLES H. EDEN.
With Map and Frontispiece. Crown 8vo. price 9s.

RAMBLES, by PATRICIUS WALKER. Reprinted from *Fraser's Magazine*,
with a Vignette of the Queen's Bower in the New Forest. Crown 8vo. 10s. 6d.

The CRUISE of HER MAJESTY'S SHIP, the CURAÇOA, AMONG
the SOUTH SEA ISLANDS in 1865. By JULIUS BRENCHLEY, Esq. M.A.
F.R.G.S. With Chart, 43 Coloured Plates and numerous other Illustrations.
Imperial 8vo. price 42s.

GUIDE to the PYRENEES, for the use of Mountaineers. By
CHARLES PACKE. With Map and Illustrations. Crown 8vo. 7s. 6d.

The ALPINE GUIDE. By JOHN BALL, M.R.I.A. late President of
the Alpine Club. 3 vols. post 8vo. Thoroughly Revised Editions, with Maps
and Illustrations:—I. *Western Alps*, 6s. 6d. II. *Central Alps*, 7s. 6d. III.
Eastern Alps, 10s. 6d.

Introduction on Alpine Travelling in General, and on the Geology
of the Alps, price 1s. Each of the Three Volumes or Parts of the *Alpine Guide*
may be had with the INTRODUCTION prefixed, price 1s. extra.

VISITS to REMARKABLE PLACES: Old Halls, Battle-Fields, and
Stones Illustrative of Striking Passages in English History and Poetry. By
WILLIAM HOWITT. 2 vols. square crown 8vo. with Woodcuts, 25s.

The RURAL LIFE of ENGLAND. By the same Author. With
Woodcuts by Bewick and Williams. Medium 8vo. 12s. 6d.

Works of Fiction.

ELENA; an Italian Tale. By L. N. COMYN, Author of ' Atherstone
Priory.' 2 vols. post 8vo. 14s.

LADY WILLOUGHBY'S DIARY, 1635—1663; Charles the First, the Protectorate, and the Restoration. Reproduced in the Style of the Period to which the Diary relates. Crown 8vo. price 7s 6d.

POPULAR ROMANCES of the MIDDLE AGES. By GEORGE W. Cox, M.A., Author of the 'Mythology of the Aryan Nations' &c. and EUSTACE HINTON JONES. Crown 8vo. price 10s. 6d.

TALES of the TEUTONIC LANDS; a Sequel to 'Popular Romances of the Middle Ages.' By the same Authors. Crown 8vo. 10s. 6d.

The FOLK-LORE of ROME, collected by Word of Mouth from the People. By R. H. BUSK, Author of 'Patrañas,' &c. Crown 8vo. 12s. 6d.

The BURGOMASTER'S FAMILY; or, Weal and Woe in a Little World. By CHRISTINE MULLER, Translated from the Dutch by Sir JOHN SHAW LEFEVRE, F.R.S. Crown 8vo. price 6s.

NOVELS and TALES. By the Right Hon. B. DISRAELI, M.P. Cabinet Edition, complete in Ten Volumes, crown 8vo. price £3.

LOTHAIR, 6s.	HENRIETTA TEMPLE, 6s.
CONINGSBY, 6s.	CONTARINI FLEMING, &c. 6s.
SYBIL, 6s.	ALROY, IXION, &c. 6s.
TANCRED, 6s.	The YOUNG DUKE, &c. 6s.
VENETIA, 6s.	VIVIAN GREY, 6s.

The MODERN NOVELIST'S LIBRARY. Each Work, in crown 8vo. complete in a Single Volume:—

ATHERSTONE PRIORY, 2s. boards; 2s. 6d. cloth.
MELVILLE'S GLADIATORS, 2s boards; 2s. 6d. cloth.
————— GOOD FOR NOTHING, 2s. boards; 2s. 6d. cloth.
————— HOLMBY HOUSE, 2s. boards; 2s. 6d. cloth.
————— INTERPRETER, 2s. boards; 2s. 6d. cloth.
————— KATE COVENTRY, 2s. boards; 2s. 6d. cloth.
————— QUEEN'S MARIES, 2s. boards; 2s. 6d. cloth.
————— DIGBY GRAND, 2s. boards; 2s. 6d. cloth.
————— GENERAL BOUNCE, 2s. boards; 2s. 6d. cloth.
TROLLOPE'S WARDEN, 1s. 6d. boards; 2s. cloth.
—————BARCHESTER TOWERS, 2s. boards; 2s. 6d. cloth.
BRAMLEY-MOORE'S SIX SISTERS of the VALLEYS, 2s. boards; 2s. 6d. cloth.

CABINET EDITION of STORIES and TALES by Miss SEWELL:—

AMY HERBERT, 2s. 6d.	IVORS, 2s. 6d.
GERTRUDE, 2s. 6d.	KATHARINE ASHTON, 2s. 6d.
The EARL's DAUGHTER, 2s. 6d.	MARGARET PERCIVAL, 3s. 6d.
EXPERIENCE of LIFE, 2s. 6d.	LANETON PARSONAGE, 3s. 6d.
CLEVE HALL, 2s. 6d.	URSULA, 3s. 6d.

CYLLENE; or, the Fall of Paganism. By HENRY SNEYD, M.A. University College, Oxford. 2 vols. post 8vo. price 14s.

BECKER'S GALLUS; or, Roman Scenes of the Time of Augustus : with Notes and Excursuses. New Edition. Post 8vo. 7s. 6d.

BECKER'S CHARICLES: a Tale illustrative of Private Life among the Ancient Greeks : with Notes and Excursuses. New Edition. Post 8vo. 7s. 6d.

TALES of ANCIENT GREECE. By GEORGE W. Cox, M.A. late Scholar of Trin. Coll. Oxon. Crown 8vo. price 6s. 6d.

Poetry and The Drama.

FAUST: a Dramatic Poem. By GOETHE. Translated into English Prose, with Notes, by A. HAYWARD. Eighth Edition. Fcp. 8vo. price 3s.

BALLADS and LYRICS of OLD FRANCE; with other Poems. By
A. LANG, Fellow of Merton College, Oxford. Square fcp. 8vo. price 5s.

MOORE'S IRISH MELODIES, Maclise's Edition, with 161 Steel Plates
from Original Drawings. Super-royal 8vo. 31s. 6d.

Miniature Edition of Moore's Irish Melodies, with Maclise's Designs (as above) reduced in Lithography. Imp. 16mo. 10s. 6d.

MOORE'S LALLA ROOKH. Tenniel's Edition, with 68 Wood
Engravings from Original Drawings and other Illustrations. Fcp. 4to. 21s.

SOUTHEY'S POETICAL WORKS, with the Author's last Corrections
and copyright Additions. Medium 8vo. with Portrait and Vignette, 14s.

LAYS of ANCIENT ROME; with IVRY and the ARMADA. By the
Right Hon. Lord MACAULAY. 16mo. 3s. 6d.

Lord Macaulay's Lays of Ancient Rome. With 90 Illustrations on
Wood, from the Antique, from Drawings by G. SCHARF. Fcp. 4to. 21s.

Miniature Edition of Lord Macaulay's Lays of Ancient Rome,
with the Illustrations (as above) reduced in Lithography. Imp. 16mo. 10s. 6d.

GOLDSMITH'S POETICAL WORKS, with Wood Engravings from
Designs by Members of the Etching-Club. Imp. 16mo. 7s. 6d.

The ÆNEID of VIRGIL Translated into English Verse. By JOHN
CONINGTON, M.A. New Edition. Crown 8vo. 9s.

The ODES and EPODES of HORACE; a Metrical Translation into
English, with Introduction and Commentaries. By Lord LYTTON. With Latin
Text. New Edition. Post 8vo. price 10s. 6d.

HORATII OPERA. Library Edition, with Marginal References and
English Notes. Edited by the Rev. J. E. YONGE. 8vo. 21s.

The LYCIDAS and EPITAPHIUM DAMONIS of MILTON. Edited,
with Notes and Introduction, by C. S. JERRAM, M.A. Trin. Coll. Oxford; including a Reprint of the rare Latin Version by W. Hogg, 1694. [In the press.

BOWDLER'S FAMILY SHAKSPEARE, cheaper Genuine Editions.
Medium 8vo. large type, with 36 WOODCUTS, price 14s. Cabinet Edition, with
the same ILLUSTRATIONS, 6 vols. fcp. 8vo. price 21s.

POEMS. By JEAN INGELOW. 2 vols. fcp. 8vo. price 10s.
FIRST SERIES, containing 'DIVIDED,' 'The STAR'S MONUMENT,' &c. Sixteenth
Thousand. Fcp. 8vo. price 5s.
SECOND SERIES, 'A STORY of DOOM,' 'GLADYS and her ISLAND,' &c. Fifth
Thousand. Fcp. 8vo. price 5s.

POEMS by Jean Ingelow. FIRST SERIES, with nearly 100 Illustrations,
engraved on Wood by Dalziel Brothers. Fcp. 4to. 21s.

Rural Sports, &c.

The DEAD SHOT; or, Sportsman's Complete Guide: a Treatise on
the Use of the Gun, Dog-breaking, Pigeon-shooting, &c. By MARKSMAN.
Revised Edition. Fcp. 8vo. with Plates, 5s.

ENCYCLOPÆDIA of RURAL SPORTS; a complete Account, Historical, Practical, and Descriptive, of Hunting, Shooting, Fishing, Racing, and all other Rural and Athletic Sports and Pastimes. By D. P. BLAINE. With above 600 Woodcuts (20 from Designs by JOHN LEECH). 8vo. 21s.

The FLY-FISHER'S ENTOMOLOGY. By ALFRED RONALDS. With coloured Representations of the Natural and Artificial Insect. Sixth Edition, with 20 coloured Plates. 8vo. 14s.

A BOOK on ANGLING; a complete Treatise on the Art of Angling in every branch. By FRANCIS FRANCIS. New Edition, with Portrait and 15 other Plates, plain and coloured. Post 8vo. 15s.

WILCOCKS'S SEA-FISHERMAN; comprising the Chief Methods of Hook and Line Fishing in the British and other Seas, a Glance at Nets, and Remarks on Boats and Boating. Second Edition, with 80 Woodcuts. Post 8vo. 12s. 6d.

HORSES and STABLES. By Colonel F. FITZWYGRAM, XV. the King's Hussars. With Twenty-four Plates of Illustrations, containing very numerous Figures engraved on Wood. 8vo. 15s.

The HORSE'S FOOT, and HOW to KEEP it SOUND. By W. MILES, Esq. Ninth Edition, with Illustrations. Imperial 8vo. 12s. 6d.

A PLAIN TREATISE on HORSE-SHOEING. By the same Author. Sixth Edition. Post 8vo. with Illustrations, 2s. 6d.

STABLES and STABLE-FITTINGS. By the same. Imp. 8vo. with 13 Plates, 15s.

REMARKS on HORSES' TEETH, addressed to Purchasers. By the same. Post 8vo. 1s. 6d.

A TREATISE on HORSE-SHOEING and LAMENESS. By JOSEPH GAMGEE, Veterinary Surgeon. 8vo. with 55 Woodcuts, price 15s.

The HORSE: with a Treatise on Draught. By WILLIAM YOUATT. New Edition, revised and enlarged. 8vo. with numerous Woodcuts, 12s. 6d.

The DOG. By the same Author. 8vo. with numerous Woodcuts, 6s.

The DOG in HEALTH and DISEASE. By STONEHENGE. With 7 Wood Engravings. Square crown 8vo. 7s. 6d.

The GREYHOUND. By STONEHENGE. Revised Edition, with 24 Portraits of Greyhounds. Square crown 8vo. 6d.

The SETTER: with Notices of the most Eminent Breeds now Extant. Instructions how to Breed, Rear, and Break, Dog Shows, Field Trials, and General Management, &c. By E. LAVERACK. Crown 4 to. with 2 plates, 7s. 6d.

The OX; his Diseases and their Treatment: with an Essay on Parturition in the Cow. By J. R. DOBSON. Crown 8vo. with Illustrations, 7s. 6d.

Works of *Utility* and *General Information.*

The THEORY and PRACTICE of BANKING. By H. D. MACLEOD, M.A. Barrister-at-Law. Second Edition, entirely remodelled. 2 vols. 8vo. 30s.

A DICTIONARY, Practical, Theoretical, and Historical, of Commerce and Commercial Navigation. By J. R. M'CULLOCH. New and thoroughly revised Edition. 8vo. price 63s. cloth, or 70s. half-bd. in russia.

The CABINET LAWYER; a Popular Digest of the Laws of England, Civil, Criminal, and Constitutional: intended for Practical Use and General Information. Twenty-fourth Edition. Fcp. 8vo. price 9s.

A PROFITABLE BOOK UPON DOMESTIC LAW; Essays for English Women and Law Students. By PERKINS, Junior, M.A. Barrister-at-Law. Post 8vo. 10s. 6d.

BLACKSTONE ECONOMISED, a Compendium of the Laws of England to the Present time, in Four Books, each embracing the Legal Principles and Practical Information contained in their respective volumes of Blackstone, supplemented by Subsequent Statutory Enactments, Important Legal Decisions, &c. By D. M. AIRD, Barrister-at-Law. Second Edition. Post 8vo. 7s. 6d.

PEWTNER'S COMPREHENSIVE SPECIFIER; a Guide to the Practical Specification of every kind of Building-Artificers' Work, with Forms of Conditions and Agreements. Edited by W. YOUNG. Crown 8vo. 6s.

COLLIERIES and COLLIERS; a Handbook of the Law and Leading Cases relating thereto. By J. C. FOWLER, of the Inner Temple, Barrister. Third Edition. Fcp. 8vo. 7s. 6d.

The MATERNAL MANAGEMENT of CHILDREN in HEALTH and Disease. By THOMAS BULL, M.D. Fcp. 5s.

HINTS to MOTHERS on the MANAGEMENT of their HEALTH during the Period of Pregnancy and in the Lying-in Room. By the late THOMAS BULL, M.D. Fcp. 5s.

The THEORY of the MODERN SCIENTIFIC GAME of WHIST. By WILLIAM POLE, F.R.S. Fifth Edition, enlarged. Fcp. 8vo. 2s. 6d.

CHESS OPENINGS. By F. W. LONGMAN, Balliol College, Oxford. Second Edition revised. Fcp. 8vo. 2s. 6d.

THREE HUNDRED ORIGINAL CHESS PROBLEMS and STUDIES. By JAMES PIERCE, M.A. and W. T. PIERCE. With numerous Diagrams. Square fcp. 8vo. 7s. 6d.

A PRACTICAL TREATISE on BREWING; with Formulæ for Public Brewers, and Instructions for Private Families. By W. BLACK. 8vo. 10s. 6d.

MODERN COOKERY for PRIVATE FAMILIES, reduced to a System of Easy Practice in a Series of carefully-tested Receipts. By ELIZA ACTON. Newly revised and enlarged; with 8 Plates and 150 Woodcuts. Fcp. 8vo. 6s.

WILLICH'S POPULAR TABLES, for ascertaining, according to the Carlisle Table of Mortality, the value of Lifehold, Leasehold, and Church Property, Renewal Fines, Reversions, &c. Re-edited by M. MARRIOTT. Post 8vo. 10s.

MAUNDER'S TREASURY of KNOWLEDGE and LIBRARY of Reference: comprising an English Dictionary and Grammar, Universal Gazetteer, Classical Dictionary, Chronology, Law Dictionary, a synopsis of the Peerage useful Tables, &c. Revised Edition. Fcp. 8vo. 6s. cloth, or 10s. calf.

INDEX.

ACTON's Modern Cookery 28
AIRD's Blackstone Economised 28
Alpine Club Map of Switzerland 24
Alpine Guide (The) 21
ALTHAUS' Medical Electricity 17
AMOS's Jurisprudence 6
——— Primer of the Constitution 6
ANDERSON's Strength of Materials........... 13
ARMSTRONG's Organic Chemistry............. 13
ARNOLD's Manual of English Literature .. 8
ARNOULD's Life of Denman 4
Atherstone Priory 25
Authority and Conscience 22
Autumn Holidays of a Country Parson 9
AYRE's Treasury of Bible Knowledge 22

BACON's Essays, by WHATELY 6
——— Life and Letters, by SPEDDING 6
——— Works, edited by SPEDDING 6
BAIN's Logic, Deductive and Inductive..... 10
——— Mental and Moral Science 10
——— on the Senses and Intellect 10
BALL's Alpine Guide 24
BAYLDON's Rents and Tillages 20
BECKER's Charicles and Gallus 25
BENFEY's Sanskrit Dictionary 8
BLACK's Treatise on Brewing 28
BLACKLEY's German-English Dictionary... 9
BLAINE's Rural Sports............................. 27
BLOXAM's Metals 13
BOASE & COURTNEY's Bibliotheca Cornu-
   biensis 4
BOULTBEE on 39 Articles 21
BOURNE's Catechism of the Steam Engine . 20
——— Handbook of Steam Engine 20
——— Improvements in the Steam
   Engine.. 20
——— Treatise on the Steam Engine ... 20
BOWDLER's Family SHAKSPEARE 26
BRAMLEY-MOORE's Six Sisters of the
   Valleys.. 25
BRANDE's Dictionary of Science, Litera-
   ture, and Art.................................... 16
BRAY's Manual of Anthropology 11
——— Philosophy of Necessity 10
——— on Force 10
BRENCHLEY's Cruise of H.M.S. Curaçoa .. 24
BRINKLEY's ASTRONOMY 11
BROWNE's Exposition of the 39 Articles..... 21
BRUNEL's Life of BRUNEL 5
BUCKLE's History of Civilization 3
——— Miscellaneous Writings............... 9
BULL's Hints to Mothers 28
——— Maternal Management of Children 28
BUNSEN's Prayers 21
Burgomaster's Family (The) 25
BURKE's Rise of Great Families................ 5
——— Vicissitudes of Families............. 5
BURTON's Christian Church 3
BUSK's Folk-Lore of Rome 24

Cabinet Lawyer 28
CAMPBELL's Norway 24
CATES's Biographical Dictionary 5
——— and WOODWARD's Encyclopædia 4
CATS' and FARLIE's Moral Emblems 18
Changed Aspects of Unchanged Truths...... 9
CHESNEY's Indian Polity 3
——— Modern Military Biography ... 4
——— Waterloo Campaign.................. 2
Christ the Consoler............................... 21
CLOUGH's Lives from Plutarch 9
CODRINGTON's (Admiral) Memoir 5
COLENSO (Bishop) on Pentateuch 22
——— on Moabite Stone, &c. 22
——— on Speaker's Bible Commentary 21
COLLINS's Perspective........................... 19
COLOMB's Slave Catching 23
Commonplace Philosopher, by A.K.H.B. ... 9
COMYN's Elena 24
CONGREVE's Politics of Aristotle 6
CONINGTON's Translation of the Æneid... 26
——— Miscellaneous Writings 9
CONTANSEAU's French-English Diction-
   aries 8
CONYBEARE and HOWSON's St. Paul 1
COOKE's Grotesque Animals.................... 18
COPLAND's Dictionary of Practical Medicine 17
COTTON's (Bishop) Memoir 5
Counsel and Comfort from a City Pulpit...... 9
COX's Aryan Mythology.......................... 3
——— Crusades................................. 4
——— History of Greece..................... 2
——— Tale of the Great Persian War 2
——— Tales of Ancient Greece.................. 25
COX and JONES's Popular Romances....... 25
——— Tales of Teutonic Lands 25
CRAWLEY's Thucydides.......................... 3
CREASY on British Constitutions 3
CRESY's Encyclopædia of Civil Engineer-
   ing 19
Critical Essays of a Country Parson 9
CROOKE's Chemical Analysis 16
——— Dyeing and Calico Printing 16
CULLEY's Handbook of Telegraphy........... 19
CUSACK's History of Ireland 3

DAVIDSON's Introduction to New Testament 22
Dead Shot (The), by MARKSMAN 26
DECAISNE and LE MAOUT's Botany 15
DE MORGAN's Budget of Paradoxes......... 11
DISRAELI's Lord George Bentinck 5
——— Novels and Tales 25
DOBSON on the Ox................................ 27
DONKIN on Diabetes 17
DOVE on Storms 12
DOYLE's Fairyland 18
DREW's Reasons of Faith 21

EASTLAKE's Hints on Household Taste...... 19
——— Gothic Revival 19

EDEN'S Queensland 24
EDWARDS'S Travels in Tyrol 23
Elements of Botany 15
ELLICOTT'S Commentary on Ephesians 22
————————————— Galatians 22
————————————— Pastoral Epist. 22
————————————— Philippians, &c 22
————————————— Thessalonians 22
———————— Lectures on the Life of Christ... 22
Epochs of History 4
ERICHSEN'S Surgery 16
EVANS'S Ancient Stone Implements............. 14
EWALD'S Antiquities of Israel 21
———————— History of Israel 22

FAIRBAIRN'S Applications of Iron 20
———————— Information for Engineers ... 20
———————— Mills and Millwork 20
FARRAR'S Chapters on Language................ 7
———————— Families of Speech 7
FITZWYGRAM on Horses and Stables 27
FORSYTH'S Essays.................................. 6
FOWLER'S Collieries and Colliers 28
FRANCIS'S Fishing Book 27
FREEMAN'S Historical Geography of Europe 4
FRESHFIELD'S Travels in the Caucasus...... 24
From January to December 15
FROUDE'S English in Ireland 1
———————— History of England 1
———————— Short Studies on Great Subjects 9

GAMGEE on Horse-Shoeing 27
GANOT'S Elementary Physics 13
————————Natural Philosophy 13
GARDINER'S Thirty Years' War 4
GARROD'S Materia Medica 18
GILBERT'S Cadore, or Titian's Country 23
GILBERT and CHURCHILL'S Dolomites...... 23
GIRDLESTONE'S Bible Synonymes............. 21
GOETHE'S Faust, translated by Hayward ... 25
GOLDSMITH'S Poems, Illustrated 26
GOODEVE'S Mechanism 13
———————— Mechanics.......................... 13
GRANT'S Ethics of Aristotle 6
Graver Thoughts of a Country Parson 9
GRAY'S Anatomy 17
GRIFFIN'S Algebra and Trigonometry 13
GRIFFITH'S Fundamentals 21
———————— Sermons for the Times........... 21
GROVE on Correlation of Physical Forces ... 14
GUYOT'S Earth and Man 14
GWILT'S Encyclopædia of Architecture...... 19

HARDING'S Texts and Thoughts 22
HARE on Election of Representatives 7
HARTWIG'S Harmonies of Nature.............. 15
———————— Polar World 15
———————— Sea and its Living Wonders ... 15
———————— Subterranean World 15
———————— Tropical World 15
HAUGHTON'S Animal Mechanics 14
HAYWARD'S Biographical and Critical
     Essays; Second and Third Series 4

HELMHOTZ'S Popular Lectures 13
HEMSLEY'S Handbook of Trees and Plants 15
HERSCHEL'S Outlines of Astronomy 11
HODGSON'S Theory of Practice 10
———————— Time and Space................... 10
HOLLAND'S Recollections 5
HOLMES'S System of Surgery 16
———————— Surgical Diseases of Infancy..... 16
HOWITT'S Rural Life of England 21
———————— Visits to Remarkable Places..... 24
HÜBNER'S Memoir of Sixtus V. 2
HUGHES'S (W.) Manual of Geography 12
HUMBOLDT'S Centenary Biography 5
HUME'S Essays 11
———————— Treatise on Human Nature 11

IHNE'S Roman History 2
INGELOW'S Poems 26

JAMESON'S Saints and Martyrs 18
———————— Legends of the Madonna 18
———————— Monastic Orders.................. 18
JAMESON and EASTLAKE'S Saviour 18
JENKIN'S Electricity and Magnetism 13
JERRAM'S Lycidas of Milton 26
JERROLD'S Life of Napoleon 4
JOHNSTON'S Geographical Dictionary......... 12

KALISCH'S Commentary on the Bible 8
KEITH on Fulfilment of Prophecy.............. 22
KENYON, Life of the First Lord 5
KERL'S Metallurgy 20
KIRBY and SPENCE'S Entomology............. 15

LANG'S Ballads and Lyrics 26
LATHAM'S English Dictionary 7
LAUGHTON'S Nautical Surveying 12
LAVERACK'S Setter 27
LAWRENCE on Rocks 14
LECKY'S History of European Morals......... 3
———————— Rationalism 3
———————— Leaders of Public Opinion 5
Leisure Hours in Town, by A.K.H.B. 9
Lessons of Middle Age, by A.K.H.B. 9
LESTER'S Atlantic to Pacific 23
LEWES' History of Philosophy 3
LIDDELL and SCOTT'S Two Lexicons 8
Life of Man Symbolised 18
LINDLEY and MOORE'S Treasury of Botany 15
LLOYD'S Wave-Theory of Light 13
LONGMAN'S Edward the Third 2
———————— Lectures on History of England 2
———————— Old and New St. Paul's 18
———————— Chess Openings 28
LOUDON'S Agriculture............................ 20
———————— Gardening 20
———————— Plants 16
LOWNDES' Engineer's Handbook 19
LUBBOCK on Origin of Civilisation 14
Lyra Germanica.............................. 18, 23
LYTTON'S Odes of Horace 26

MACAULAY'S (Lord) Essays 1
———————————History of England ... 3
———————————Lays of Ancient Rome 26

MACAULAY'S (Lord) Miscellaneous Writings 9
———————Speeches 7
———————Complete Works......... 1
MACLEOD'S Economical Philosophy .,....... 7
——————— Theory and Practice of Banking 27
McCULLOCH'S Dictionary of Commerce ... 28
MARKHAM'S History of Persia 3
MARSHALL'S Physiology............................. 17
——————— Todas 11
MARSHMAN'S Life of Havelock 5
——————— History of India 2
MARTINEAU'S Christian Life 23
——————— Hymns 23
MAUNDER'S Biographical Treasury 6
——————— Geographical Treasury........... 12
——————— Historical Treasury 3
——————— Scientific and Literary Trea-
sury 16
——————— Treasury of Knowledge............ 28
——————— Treasury of Natural History .. 15
MAXWELL'S Theory of Heat...................... 13
MAY'S Constitutional History of England.. 1
——— History of Democracy 1
MELVILLE'S Novels and Tales 25
MENDELSSOHN'S Letters 5
MERIVALE'S Fall of the Roman Republic... 2
——————— Romans under the Empire ... 2
MERRIFIELD'S Arithmetic & Mensuration . 13
——————— Magnetism 12
MILES on Horse's Feet and Horseshoeing ... 27
——————— Horses' Teeth and Stables........... 27
MILL (J.) on the Mind.............................. 10
MILL (J. S.) on Liberty 7
——————— on Representative Government 7
——————— on Utilitarianism 7
——————'s (J.S.) Autobiography............... 4
——————— Dissertations and Discussions 7
——————— Political Economy 7
——————— System of Logic 7
——————— Hamilton's Philosophy 7
——————— Subjection of Women 7
——————— Unsettled Questions 7
MILLER'S Elements of Chemistry 16
——————— Inorganic Chemistry 13
MINTO'S (Lord) Life and Letters 4
MITCHELL'S Manual of Architecture 19
——————— Manual of Assaying 20
MONSELL'S Spiritual Songs 23
MOORE'S Irish Melodies.......................... 26
——————— Lalla Rookh 26
MORELL'S Elements of Psychology 10
——————— Mental Philosophy 10
MORRIS'S French Revolution.................... 4
MOSSMAN'S Catholic Church 3
MÜLLER'S (MAX) Chips from a German
Workshop 10
——————— Lectures on Language 8
——————— Science of Religion 20
MURCHISON on Continued Fevers............. 17
——————— on Liver Complaints 17

New Testament, Illustrated Edition........... 18
NORTHCOTT'S Lathes and Turning 19

O'CONON'S Commentary on Hebrews 21

O'CONOR'S Commentary on Romans 21
ODLING'S Course of Practical Chemistry ... 15
O'Keeffe Case (the) 11
OWEN'S Lectures on the Invertebrata 14
——————— Comparative Anatomy and Physio-
logy of Vertebrate Animals ... 14

PACKE'S Guide to the Pyrenees 24
PAGET'S Lectures on Surgical Pathology ... 17
PAYEN'S Industrial Chemistry 19
PEREIRA'S Elements of Materia Medica ... 17
PERKINS'S Legal Essays........................... 28
PETIT'S History of Mary Stuart................. 2
PEWTNER'S Comprehensive Specifier 28
PIERCE'S Chess Problems 28
POLE on Whist 28
PRENDERGAST'S Mastery of Languages 9
Present-Day Thoughts, by A. K. H. B. 9
PROCTOR'S Astronomical Essays 11
——————— Moon 11
——————— New Star Atlas 12
——————— Orbs Around Us................. 12
——————— Plurality of Worlds 11
——————— Saturn and its System......... 12
——————— Scientific Essays (Two Series)... 14
——————— Sun 11
——————— Universe 11
Public Schools Atlases (The) 12
——————— Modern Geography............... 12

QUAIN'S Anatomy.................................... 17

RAMSAY'S Geology for Beginners 4
RANKEN on Strains in Trusses 19
RAWLINSON'S Parthia............................... 2
——————— Sassanian Monarchy 2
Recreations of a Country Parson 9
REDGRAVE'S Dictionary of Artists 18
REILLY'S Map of Mont Blanc 24
RICH'S Dictionary of Antiquities 8
RIVERS' Rose Amateur's Guide 15
ROGERS'S Eclipse of Faith...................... 10
——————— Defence of ditto 10
ROGET'S English Words and Phrases 7
RONALD'S Fly-Fisher's Entomology 27
ROTHSCHILD'S Israelites........................... 22
RUSSELL'S (Count) Pau and the Pyrenees... 24
RUSSELL (Lord) on Christian Religion...... 29
——————— on Constitution & Government 1

SALTER on the Teeth 17
SANDARS'S Justinian Institutes 6
SANFORD'S English Kings......................... 1
SAVORY'S Geometric Turning 19
SCHELLEN'S Spectrum Analysis................. 12
SCOTT'S Albert Durer 18
Seaside Musings by A. K. H. B. 9
SEEBOHM'S Oxford Reformers of 1498 2
——————— Protestant Revolution 4
SEWELL'S Examination for Confirmation... 22
——————— History of the Early Church 3
——————— Passing Thoughts on Religion ... 22
——————— Preparations for Communion..... 23
——————— Principles of Education 23

SEWELL'S Readings for Confirmation   23
———— Readings for Lent   23
———— Tales and Stories   25
———— Thoughts for the Age..................   22
———— Thoughts for the Holy Week......   23
SHARP'S Post Office Gazetteer   12
SHELLEY'S Workshop Appliances...............   13
SHORT'S Church History............................   3
SIMPSON'S Meeting the Sun   23
SMITH's (J.) Paul's Voyage and Shipwreck   21
———— (SYDNEY) Essays   10
———————— Life and Letters   5
———————— Miscellaneous Works ...   10
———————— Wit and Wisdom   10
———— (Dr. R. A.) Air and Rain..............   12
SNEYD'S Cyllene   25
SOUTHEY'S Doctor.....................................   8
———— Poetical Works   26
STANLEY'S History of British Birds   15
STEPHEN's Ecclesiastical Biography   6
———— Freethinking & Plain Speaking   9
STIRLING'S HAMILTON   10
———— HEGEL   10
STOCKMAR'S (Baron) Memoirs....................   5
STOKES'S Lectures on Fever...........   17
STONEHENGE on the Dog   27
———— on the Greyhound...............   27
STRICKLAND'S Queens of England.............   5
Sunday Afternoons, by A. K. H. B...............   9
Supernatural Religion   20

TAYLOR'S History of India   2
———— (Jeremy) Works, edited by EDEN   23
Text-Books of Science   13
THIRLWALL'S History of Greece   2
THOMSON's Laws of Thought   7
THORPE's Quantitative Chemical Analysis   13
THORPE and MUIR's Qualitative Analysis   13
THUDICHUM's Chemical Physiology   16
TODD (A.) on Parliamentary Government...   1
TODD and BOWMAN'S Anatomy and Phy-
    siology of Man   17
TRENCH's Realities of Irish Life................   3
TROLLOPE's Barchester Towers   25
———— Warden   25
TYNDALL on Diamagnetism   13
———— Electricity   13
———— Heat   13
———— Sound   13
———— 's American Lectures on Light .   13

TYNDALL'S Fragments of Science   14
———— Hours of Exercise in the Alps   21
———— Lectures on Light   14
———— Molecular Physics...................   13

UEBERWEG'S System of Logic.....................   11
URE's Arts, Manufactures, and Mines.........   19

WALKER's Rambles   24
WALTER'S Home and Rural Life of
    Shakespeare   4
WATSON's Geometry   13
———— Principles & Practice of Physic .   17
WATTS's Dictionary of Chemistry..............   18
WEBB's Objects for Common Telescopes ...   12
WELLINGTON'S Life, by GLEIG   5
WEST on Children's Diseases   16
WHATELY's English Synonymes   6
———— Lessons on Christian Evidences   23
———— Logic.....................................   6
———— Rhetoric   6
WHITE'S St. Mark's Gospel   21
———— St. John's Gospel   22
———— Latin-English and English-Latin
    Dictionaries   8
WHITE & DONKIN's English Dictionary ...   7
WHITWORTH on Guns and Steel   20
WILCOCKS's Sea Fisherman   27
WILLIAMS's Aristotle's Ethics   6
WILLICH's Popular Tables   26
WILLIS's Principl of Mechanism   19
WILLOUGHBY's (Lady) Diary.....................   25
WOOD's Bible Animals   14
———— Homes without Hands .... ——   14
———— Insects at Home   14
———— Abroad.....................   14
———— Out of Doors   15
———— Strange Dwellings   15
WORDSWORTH'S Christian Ministry   21

YONGE's English-Greek Lexicons..............   8
———— Horace.....................................   26
———— History of England   1
———— English Literature....................   8
———— Modern History   2
YOUATT on the Dog   27
———— on the Horse   27

ZELLER'S Socrates   6
———— Stoics, Epicureans, and Sceptics .   6

www.ingramcontent.com/pod-product-compliance
Lightning Source LLC
Chambersburg PA
CBHW021354210326
41599CB00011B/869